FINAL PROOF

By the Same Author

THE WEAK AND THE STRONG
DREAMING SPIRES (POETRY)

FINAL PROOF

JULIA SAVARESE

W · W · NORTON & COMPANY · INC ·

NEW YORK

First Edition

Library of Congress Catalog Card No. 76-116110

SBN 393 08618 6

Published simultaneously in Canada
by George J. McLeod Limited, Toronto

Designed by Ruth Smerechniak

PRINTED IN THE UNITED STATES OF AMERICA

1 2 3 4 5 6 7 8 9 0

FOR

Louise and Bill Phillips

Harry Sions

Merrill Pollack

William Evans

Nathan Somberg

Eric Cassell

Each of whom loved and labored to "roll away the stone" in the face of the strongest and most stubborn opposition: me.

With love and gratitude.

New York City, 1970

FINAL PROOF

one

"I CAN'T STAND TO BE IN BED WITHOUT I hev somethin' warm to tetch." Patty's oval face looked like something yearning to cry, but she forced the sound so it came out gay and a little funny. "I go pure out o' my mind was I to wake up and there be nothin' there but me and the dark . . ."

I didn't know what to say so I looked down toward the street. New York was blunted with cold that winter evening; from my window I could see a clot of people waiting for the uptown Madison Avenue bus.

Patty and I were among the last ones still at work. I got heartsick every time I thought about Patty, much less spoke to her. Blonde, with real cornflower blue eyes, she was at nineteen as completely innocent as a child while more than halfway home to being a whore. I remembered the time I had heard someone in the ladies' room refer to her as "the sleeping bag of the Kimberly

Publishing Company." My stomach had tightened like a fist—the same way it did now with my remembering it.

She had come straight to New York from Georgia, managing to get screwed by a friendly truck driver who had stopped on the road to give her a lift. She sent half her pay from being a secretary at Kimberly back home to feed the five kids that were still there, and she had slept with just about all the brass in the company just so she wouldn't have to sleep alone.

"I come up to New York 'cause Mister Rusk . . . I meet him down homeways . . . in Atlanta, you know. It was at a kind of business meetin' . . . a sales meetin' I think they call it . . . mostly men it was . . . and the las' night of the meetin' they had a big party . . . and they invited in some of the girls in town . . . you know, like to make it less like workin' than they'd been doin'. That was then I met up with Mister Rusk. And later on that same night he tell me, 'You ever want to come up to New York, I set it up so you have a good job at this big publishin' company I'm with—the Kimberly Publishing Company.' And so I come up." Mostly she slept with either Rusk or one of his henchmen; officially she was secretary to a couple of Kimberly salesmen who sold space for the magazines.

"You know what?" she said slowly, "sometimes I get the feelin' that Mister Rusk didn't really 'spect me to come . . . Well," she shrugged off the idea, "I guess I better get scootin'—I'm to meet a Mister Mills at the Drake Hotel Bar . . . Mister Rusk's a friend of his. I never met him as yet—this Mister Mills, that is. But Mister Rusk said I'd like him. He works for Kimberly too—some kind of executive, I forget exactly what. Well, I guess I be going." She started out of the room and then stopped. "Oh, honey, you got a stamp maybe? I got to get this letter off to home. My brother, Jerry—he's the littlest one—he had himself a bad attack of the flu. I bought a funny card for him to look at. It's Snoopy, ya know?"

I found a stamp in my bag and took the six cents she insisted on giving to me; then I watched her leave for the Drake Hotel Bar and Mister Mills.

I thought of Rusk, out in Georgia, taking this kid and playing it big. I could feel myself getting more and more furious, and I remembered a comment I had read in somebody's column:

"Kimberly is as riddled with sex as a Swiss cheese is with holes."

"What the hell," I thought, shrugging, trying not think what it might be like being nineteen and coming from a family where you slept with seven other people in the same shack—how, if you had done that, it might be more than possible that you'd go "pure out o' your mind was there to be nothin' there but you and the dark . . ."

It was a dangerous thought because it reminded me of Steve, and up to then I had managed not to think of him once the whole afternoon. For a second the remembered pain swerved like a knife across my chest. Anything was better than thinking about Steve. Even getting sick to my stomach with what was happening to Patty. The hell with it—did everything you thought about always have to *hurt?*

To distract myself I turned back to the vintage Remington. The sheet of bond hanging limply from the typewriter still looked like a fat, white tongue sticking itself out at me. And not without reason, since after an hour all I had down was, "Dear Mr. Canfield . . ."

"Go home," I said to myself. "Kimberly's done without your help for seventy years. Any reason why they're going to fold up if you desert them now?"

It was a straw-man argument if ever there was one, and I didn't waste a second before smugly knocking it down. Because—with or without my help—that's exactly what the Kimberly Publishing Company was doing. Not that I understood much about *how* it was all fouled up. I hadn't made much sense out of an item in a gossip column about some Madison Avenue advertising agency executive who had jazzed into the Commodore bar waving a ten-dollar bill, announcing that he'd take all bets that Kimberly would be dead before spring. All I did know was that any company that had fired seven people with whom I was personally acquainted on the day before Thanksgiving and that had a baby-fat-at-forty policy coordinator named Brad Rusk who walked up and down the halls yelling "Fuck *The New Yorker!*" had to be doing something wrong.

It was almost (but not quite) a joke to think I had come here right after my novel had been published (with reviews that even people not in my own family considered pretty good) merely to take a job . . . "any job" that would sustain me economically and emotionally while I continued with my Real Work.

What I had in mind was something that would break the pattern of always working alone from eight in the morning till nine at night (usually in my pajamas), typing in winter through woolen gloves, with a blanket wrapped around me *and* the little plug-in stove (since the *palazzo* we lived in had steam heat that never seemed to make it as far as the front room where I worked) and my mother creeping in every once in a while to slip me trays of Campbell's Tomato Soup and Lipton Tea with wedges of Kraft Velveeta on melba toast.

If the typewriter was going, she wouldn't say a word. But if I "wasn't doing anything"—her description for whatever might be going on in my head when the keys weren't moving (and usually right)—she would put down the tray and lean against the doorway and give me the news of the day.

Sometimes they were just weather forecasts. I think she thought of me as some exile to whom she brought "bulletins" of what was happening in the outside world. But, no matter what the news might be about—Jackie, Zsa Zsa or Julie—she always ended up her little visits with a smile and the words, "Holy God, how proud Jerry would have been of your being a writer and everything . . ."

I used to make fun of her saying that—the warm visions of my father, who had died when I was seven, each year fading dimmer and more irrevocably away—but there were days when I waited for those words with as much hunger as I did the tomato soup and Kraft Velveeta.

Since making the decision that I had to do something that would get me out of the house (and my pajamas), I had amassed six short stories, of which three were published; four outlines of a book, none of which had gotten past the third chapter; and (for whatever it was worth) seven years of experience at the Kimberly Publishing Company.

The slow-up hadn't happened overnight. In the beginning the schedule had been exactly according to plan. The job I had was to type all the copy that came in from the department editors on Kimberly's big home-service magazine, *Woman's World*. In that first week I typed everything from "Why the RH Factor Can Be a Blessing" to "Hot Dog! Every Wiener a Winner." For the first year the system worked without a hitch. Nine-to-five for Kimberly; the rest for me. If there were any publishing-type machinations going on around me, I was as ignorant of them as if I'd been putting in time in a day nursery. My lunch hours were spent at the nearest branch of the Public Library, or back at my desk digesting whatever book I had taken out (I was partial to poetry and sex crimes) together with my brought-from-home American cheese on Duggan's Whole Wheat.

And then, one day, on impulse, I had sent a note to the then newly appointed editor of *Woman's World* (a nice, middle-aged man who had been a copy editor and whom I used to run into every once in a while at the library). Even now—six years later—I could remember the exact words that had thrown the switch: "Congratulations! If I can be used in any more involved way than I am at the moment, I would like to offer my services . . ." It wasn't as though anybody had *pushed* me into anything . . .

But now—aware of it getting late—I knew that my beginnings at Kimberly weren't what I wanted to think about right now. What I wanted to do was get on with the letter I had started to Jim Canfield, the editor of *Today*, the Kimberly magazine I had been with for four months now—a letter I had been trying to write ever since I had heard that afternoon that he had been taken to the hospital after a heart attack. For a kid who was being paid to write "promotion"—a term that still didn't make *too* much sense to me—I was having serious trouble with a simple letter. For some reason I kept getting hung up on how much I didn't belong in this job, even though I guess I should have been grateful, since I had gotten it after several months of serving as a confirming statistic to the current Madison Avenue quip that there were only two kinds of Kimberly employees, "those who have been fired and those who are going to be."

I started trying to figure out what had practically indentured me to this job during all of the seven years. In my mind I ran through the times I had thought about leaving . . . times when, at least for a day, the desire to get back to my own writing full time was as strong as sex . . . times when complications seemed ridiculous and I thought of just going to whatever company was in the next building and asking for the first job they had open . . .

Impulsively I yanked the "letter" out of the typewriter and jammed in another piece of paper. Then I pulled open the desk drawer and found a clipping I had torn out of the Sunday *Times*—"Typists Wanted for the New York Telephone Company. Simple, pleasant work, congenial surroundings . . ."

Writing this one was a snap . . .

"Dear Sir: I would like to apply for one of the typist openings mentioned in this Sunday's *Times.* I am twenty-seven years old, have worked for the Kimberly Publishing Company for the past seven years and am——"

Banging away at the typewriter, envisioning the peace and "congenial surroundings" that awaited me at the New York Telephone Company, I was startled by the southern-rasp voice of Tom Redford right behind me saying, "You want to leave off mauling at that typewriter for a minute, honey, and get yourself introduced to some important people?"

I swung around, reacting with equal parts of pleasure and suspicion—my standard reflex toward Tom Redford, publisher of *Today.* Ever since the time Mike Nivera, who had been president of Kimberly, had suggested my working for *Today,* Redford had been a sort of sarcastic mentor. He was short but as straight as a Ku Klux cross, with a lock of thick blond hair which he was given to tossing back with a haughty gesture, and a smile that was as insinuating as it was charming, like somebody managing to rub against your breast while he held your elbow helping you across the street. He was also, incidentally, my first up-close Mississippian.

From the beginning Tom Redford had, in his own southernsly way, made things both possible and difficult for me. Possible because, as I said before, four months ago I didn't even know what the word "promotion" meant, and being hired by Redford instead of by the man who ran the Promotion Department, one, gave everybody at *Today* the very distinct impression that all was not just a publisher-promotion writer relationship between us, and, two, gave me the chance to find out how to do whatever it was I was supposed to do before they found *me* out.

When I first spoke to him and told him I didn't know anything about the job he was offering me, he said, "Shit, honey . . . I don't want any high-powered type promotion lady. Why with your kind of good-looking Eyetalian-type face and that cool but sexy figure, what you gonna do is bring a little *class* around this place . . ."

Now he hovered over my desk, all five-foot six and a half, honey-smooth elegance of him, and I smiled in stiff coolness as Tom Redford introduced me to Brad Rusk, the policy coordinator of the magazines, and Dave Wolffe, a tall, dark, tight-muscled man with Jewish-sad eyes who had come into the company several months before with enough rumors trailing from him to make miles of political confetti. The rumors ranged all the way from his having been involved with some fiasco in a high-powered construction firm to his having been brought in by the bankers to liquidate Kimberly. Unlike my feelings toward Rusk, which were pretty solidly unfavorable, up to now I didn't have anything in particular against Wolffe—outside of just not liking him. Now as I saw him peering down at the paper that was stuck into my typewriter my blood reached an instant boil. I started to say something, but Redford, sly as a snake when he wanted to be, saw what was going on and said, "Ain't she something, working after everybody else's gone off? You ain't maybe thinking of trying for my job *already*, are you, honey?" He went to put his arm around my shoulders, but it slipped off and flopped down at his side as I yanked the paper furiously out of the machine.

"What is it you do *exactly*, Miss Paula Jericho?" Dave Wolffe asked me in a quiet, cool voice, his sharp, bright eyes mocking as well as serious, so that in that second I found myself thinking, without it making any sense at all: *Before this is over one of us may kill the other.*

God knows what I might have come out with, except that before I could answer him Tom Redford was saying, "Hey, don't go pinning her down like that. Miss Paula here practically runs this magazine . . . 'cept I don't want her to know it. She's getting too big for her petticoats as it is."

"Well, Tom," Brad Rusk said, shifting his weight, "we better be getting into your office," and once again, as I did almost every time I saw Rusk, I hated the whole flabby, fanatic-eyed, hulking mass of him.

It was right then that I remembered the day I had come on him in the hallway where, like a wild man, he had been furiously punching and kicking away at the Coke machine, swearing and sputtering, "Screw you, you fucking, cock-sucker, son-of-a-bitch . . . I'm the one responsible for your being here and I'm the one that can kill you with one stinking phone call, you miserable, mother-fucking cheat!" There had been honest-to-God tears of personal outrage in his eyes.

Now as they all turned stiffly and started out of the office, I felt the nuttiest need to call them back, as though I knew there was something all three of them were going to decide and there was something I had to say before they started. It was a pretty stupid thought, but one I couldn't get rid of even after they had all gone into Redford's walnut-paneled office and closed the door.

Redford was right; it was much later than I had realized, but for some reason I couldn't move myself from the desk. I was amused in a way by Redford's pretending that I had to be introduced to Brad Rusk. Even aside from the Coke machine incident (he had actually had it ripped out of the building that same afternoon) our paths had crossed more than once before at Kimberly.

The last time had been in Jim Canfield's office the first week I had come to *Today*. Jim Canfield had started as editor of *Today* thirty-five years ago when it first came out, and over the years the whole development and success of the magazine had made publishing history—plus myth.

I had heard many stories about Jim Canfield before that day when Redford sent me over to meet him. It was known—or rumored—that he was irascible, testy, brilliant, a born editor. It was also known—or rumored—that there were entire days when he did a fabulous job of being both a brilliant editor and an absolute drunk. What was absolute and public *fact* was that he alone of all the staff members had attacked Brad Rusk's appointment as policy coordinator of all of the Kimberly magazines by publicly asking in a television interview whether Rusk's plan was to ration out a half-hour a day to the publishing, the advertising, the promotion, *and* the editing of each of the various magazines.

I remembered walking into Jim Canfield's office that one and only time. Jim Canfield hadn't gotten up. The office was a big, airy room full of magazines and plaques and pictures and mementos, and it smelled of books and tobacco and rubber cement. The main thing I saw was a tall, flat-faced man with something of the Middle West about him. His shoulders were thin and straight, and you felt he resented the fact that he had aged a few years since someone had taken the snapshot on the wall of him holding a tennis racket and squinting into the sun. His eyes were flat and very pale blue and didn't betray a thing.

I had come prepared with nothing to say except maybe the usual clichés. But his first sentence, thank God, saved me from any of that.

"You got any better idea than I have why that con man Redford set up this meeting?" he asked me, his eyes steely.

"No," I said, "no, I haven't."

"O.K., then, we're even." He reached for the pack of Philip Morris on his desk. "I guess you smoke . . ."

"Yes," I said, "thank you."

"Three things my mother warned me against when I left Nebraska . . ." His voice had a kind of twang to it. "Whiskey . . . anything made with mayonnaise in the summertime . . . and women who smoke in public. That was forty-eight years ago."

It was only then, as he struggled to stand up, that I realized the difficulty he had in moving from the chair. But something stopped me from making any effort to reach for the cigarette. With careful, and obviously expensive movements as far as pain was concerned, he gave me the cigarette, held the match for me and then returned to his chair.

Once he had settled back and started his own cigarette, he looked at me for a moment and then laughed. "You didn't much like it, did you?"

I said, "I beg your pardon?"

"My not standing up when you came in. You put me down for one of those thin-blood editorial-type bastards. Am I right?"

"No, sir. I mean . . ."

"Well, maybe not in so many words. Your good Italian mother probably brought you up better than that. But you didn't like it, just the same. Am I right?"

I didn't answer and he said, "Admit it," and in those two words I suddenly realized how strong any enemies would have to be to get this old one and how they had better make sure they aimed well the first time.

"Admit it."

And I smiled, facing the fact of no escape, and said, "Yes."

"And so you should have," he said. "Damn lack of simple good manners in business today. It's no wonder the economy's shot to hell. Of course you didn't like it. First thing I admired about you, young lady . . . not taking bad manners for granted." He ran his hand impatiently through his thick white hair.

"What did Tom Redford expect that you'd do? Sing me a hymn about how you'd read *Today* since you were two years old and what a deathless privilege it was to be allowed to work for such an important magazine?"

"Something like that, I guess."

"Fool southerner." He laughed. And then he was serious. "You're not mixed up with that crazy con man, are you?" And for the first time, in the middle of all the whispering, here was somebody asking the question outright . . . somebody, furthermore, who—having met me five minutes ago—seemed to care completely about the truth of the matter. And, furthermore, to care not out of curiosity, but for what seemed my sake.

"No," I said, "I'm not mixed up."

"Good. Don't be," he said. "You know that he tries to give the impression that you are . . . you know that, don't you?"

And I said, "Why would he do that?"

He laughed and then looked at me very seriously. "I'll give you the benefit of the doubt," he said, "meaning that you're serious and not just fishing for a compliment. Because it builds him up, that's why."

"But, he's the publisher."

"And all you happen to be is a young, good-looking girl, with soft brown hair that just happens to be long enough to reach her shoulders and a kind of smoldering but cool Italian figure and face . . .

"Look," he said, "it's not that I disvalue charm. And, God knows, he could probably sell snowshoes to Zulus. It's just that for some reason whenever I'm with Tom Redford I think of a theorem of Euclid some teacher or other once pounded into me, 'A line is length without breadth.'" He laughed. "That's not altogether fair, I guess. Not when you have to work for him anyway. How come you are?" he asked me. "He mentioned something about your having been in editorial once. Was it at Kimberly?"

"*Woman's World*," I said.

"Oh, one of the ladies' aid magazines." He ran his hand through his hair. "How did you happen to switch?"

And that was how—half an hour after I walked into his office—this stranger got to know more about me than anybody I'd known in the company for seven whole years.

He didn't say a word through almost the whole story . . . about my writing . . . and how I'd come to Kimberly . . . and about my deciding I wanted a job with more to think about, and how what had happened was that after I sent my simple little "can I have something more to do" note to the editor of *Woman's World*, Fritz Mott (a sweet, skinny, giraffe of a man who wore long, licorice-thin ties and smoked fat little cigars—all of which made him look exactly like one of those dressed-up overstuffed toys they sell at F. A. O. Schwartz), he took me out to a very posh lunch at the Doral, told me he was going to need "a lot of help from everyone" in his new job and asked me if I'd like to have the job of managing editor. And so, as they say, "involvement" set in. The job was exciting, challenging, and rather seductively rich for my up-to-now writer's-thin blood. In addition to making more money, I found that I actually enjoyed being part of a steadily accelerating carousel that included meetings, expense-account lunches, conferences, and an Air Travel Card that would take me anywhere in North America—even if, for six months, I never got to use it.

Unfortunately, at the end of the sixth month, I did get to use it . . . quite often. A new project came up—a spin-off of *Woman's World* to be called "The New York Woman's World," which was to be edited just for the one state. The idea was a pet of the then-president of the company, who had since left to start his own string of magazines, Mike Nivera, a young Puerto Rican whose overwhelming vitality always scared me, not because of rank but because I was never sure in any conversation whether I was going to be able to speak fast enough to keep him listening. Anyway, I guess out of laziness as much as anything else, my boss, Fritz Mott, passed the project along to me. Even that might not have *had* to be so deadly except that, first, it turned out very well; second, Nivera was impressed by its turning out very well; third, Nivera was impressed with *me* because of how impressed he was by its turning out so well; and fourth, "Mama Mott" heard about it.

Put together "Mama Mott," whose name was Maud, and sweet, weak, bumbling, over-toilet-trained Daddy Giraffe, add the appointment of the exquisitely egocentric policy coordinator, "Baby Fat" Brad Rusk . . . and the fat was in the fire. Meaning *mine*, of course.

I never knew what particular kind of brainwashing "Mama Mott" went in for since she first got it into her head that I might *just* become some sort of threat to Papa's job; all I knew was that every morning for weeks I watched my boss come in to work so progressively office-stalking, pipe-pulling, tight-pants nervous that there were moments when it seemed that the least I could do to put the dear out of his misery was to help him fire me.

Anyway, the afternoon arrived when, armed with four martinis, the ringing words of Mama-Bear-at-Breakfast, plus the immoral support of a stick-legged witch of a lady executive editor whose lunchtime function was to serve as a kind of away-from-home Mama Bear—and the stage was set for the scene I shall call "Papa Bear and me—the Parting."

It was when I got to this part in my story that Jim Canfield laughed out loud. "You keep a pretty objective view of things, don't you? Most things anyway."

For the first time I noticed the thinness of his hands; they, more than anything else about him, seemed to betray his age. His eyes, on the other hand, were as alert as a fox's.

"So what happened then . . . after Daddy Bear did his duty to all that toilet-training Mama Maud had given him? You find yourself out in the street, did you?"

"Not exactly. I mean, well, I guess I did in a way—except I refused."

"Refused?"

"To be fired."

He laughed again. Suddenly I was aware of how much of his time I was taking up; he must have seen me glance at my watch because he said, "I've got time if you have. Of course, you may have nicer things waiting to do than I have." And there wasn't any snottiness in his saying that, either; just good, simple straight fact. I blushed and went on. "What I mean is I figured since I'd been working almost directly for Mr. Nivera on the New York thing and since he seemed to like what I'd been doing I'd go to him."

"Straight to the top man, huh?"

"Well, I'd been doing the project for him. I figured the least he could do was something for me now."

"That how you ended up at *Today?*" he said.

"Not right away. Mr. Nivera said not to worry, that I wouldn't be taken off the payroll or anything. Just to wait and he'd find a job I liked for me. Maybe even to take a vacation while I was waiting. He was really very kind to me."

"Kind enough to do everything except get your old job back for you."

I didn't answer and he laughed and said: "I like the way you don't say anything sometimes . . . not that you have to with that *appassionata* Italian face of yours. You may make your living in many strange ways before you're through, but one thing you'll never be, young lady, and that's a poker player."

It was at that moment that the door was thrown open and Brad Rusk exploded into the room. He was so angry I'm not even sure he realized that anyone besides Jim Canfield was there. Even if he had, I don't think that someone else being in the room would have been able to stop the obscene, irrational flow of his fury. Sputtering, red with rage, he stomped over to Canfield's desk, waving the back page of *The New York Times* in his hand.

"What the hell is this crap supposed to be?" he demanded.

"Offhand I'd say it was a page from a newspaper," Canfield answered quietly.

"Don't you be fucking flip with me!" Rusk screamed. " 'A Salute to Jim Canfield on Thirty-Five Years of *Today* . . .' " He spat out the words. " 'We, the undersigned members of the advertising fraternity, wish to congratulate Jim Canfield, as the founder and editor for the past thirty-five years of *Today* magazine. Considering that this is probably the first time in the history of anything—much less the publishing world—that fifteen agency presidents have ever been of one mind'—shit!" He threw the paper to the floor, an action so corny that for a second I was able to breathe again and forget how scared I was that he had gone off his rocker. But the next second I found myself just as scared as I had been before, because Rusk's sputtering had only given him added momentum. "What the hell is it supposed to mean? A full-page ad in *The New York Times*. Here's the company trying to cut down on every red cent . . . every red cent. Do you know how much a full-page ad in *The New York Times* costs?"

"No," Canfield said quietly, "no, I don't. Why don't you consult the accounting department? Better still, why don't you ask the fifteen gentlemen who evidently *paid* for the ad?"

"*Gentlemen*," Rusk exploded. "Madison Avenue henchmen. Just because they own some crummy ad agency or something, that gives them the right to decide who's a good *editor*. Gentlemen . . . pimps, that's what they are!"

"Not necessarily the right to decide," Canfield said quietly, ignoring Rusk's last remark. "Just enough money to *pay* for saying they've decided." He laughed, and it was like the last thing needed to make Rusk one hundred percent loony.

"Lousy little senile self-Valentine," Rusk screamed, bending down to pick up the paper from the floor just so he could wave it around again. "Here I am trying to save this goddamn ancient company practically single-handed . . . squeezing every penny . . . every damn last penny . . . squeezing every goddamn last penny——"

"Does that include the ones for the lingerie?" Canfield interrupted quietly.

It was as if someone had hit Rusk in the face with a damp towel stolen from the Dixie Hotel.

"Lingerie . . . what lingerie?"

"The pretty little blue ones you bought for that sassy-assed little blonde researcher of yours at Saks last Thursday. Size ten, wasn't it? But I guess you figured you had a right to splurge on something . . . working so hard at squeezing all those pennies like you've been doing——"

"Why you . . ." Rusk screamed, "you . . . *you* . . ." I started counting on the number of times he said "you"—it gave me something to do to keep my mind from taking in how really loony all this had become.

And then all of a sudden, while I was concentrating on counting, I became aware of Jim Canfield talking again. Talking steadily, but so quiet, so firm and steady, that I wasn't sure whether he'd just started or whether I hadn't even heard him begin.

"Just listen," he said, "*listen.* I've worked with words—one way or another—most of my life. Some I like . . . some I don't give a damn about . . . and some I don't understand. I guess they're the ones that bug me most . . . loose, sloppy words that can mean one out of fifteen different things—or nothing. Big, high-falutin' words so swelled out with puffery they're like balloons full of air. So that's why when you take all the effort and thought and sweat and love it takes to put together seven entirely different magazines each month—to say nothing of the publishing end of it—and you put all that in the hands of one man—one man who probably doesn't know what any *one* of those magazines is really about, and you call him policy coordinator—then I'm sorry . . . I'm sorry, but I just don't understand what that title . . . what those words are supposed to mean.

"So maybe if you'd explain the term to me, Mr. Policy Coordinator . . . it must have been one of the many things I missed over the thirty-five years I've been here."

Rusk started to speak, but he never got past the first breath. Canfield continued, "As for that newspaper you keep waving around. I'm not that much happier about it than you are. What the hell does an editor need Madison Avenue telling him he's done a good job for? God knows, if they'd asked me I'd have killed the whole idea. But now that it's done . . . well, why not let the fifteen 'big guys' get their names on the bottom of something they paid for, for a change? What I do find a damn sight more interesting—and enlightening—is your reaction to it. It wouldn't be that maybe it's going to make getting rid of me a little bit rougher than you'd planned on, is it?"

"What the hell is that supposed to mean?" Rusk managed to get in.

Canfield laughed quietly. "That's right. I forgot you 'policy coordinator' types are too creative to know about crass things like advertising agencies and that—insane as it may seem—the fact that the presidents of fifteen of those agencies expressed themselves—in print, publicly—in favor of a certain editor . . . might mean that those same fifteen men might not care to have themselves made fools of—also publicly—if that editor happened to be fired. None of which matters a bit—except for the fact that those same fifteen men just coincidentally happen to handle about seventeen billion dollars' worth of advertising billings every year, a fair amount of which they either do or do not suggest putting into Kimberly publications . . ."

"*Well* . . ." Rusk said, "well . . . I'll just speak to Dave Wolffe about this whole thing. We'll just see what Dave has to say about it." His face wasn't red anymore. It had gotten kind of biscuit-dough white by now, especially around his chin and neck.

"You do that," Jim Canfield said quietly, "you just speak to savior David, O.K.?"

Rusk was halfway out the door when Canfield spoke again. And he didn't speak loudly, either, but even, so you knew that, loud or not, Rusk didn't have one chance in twenty-four million of not stopping and paying attention to what Canfield was going to say.

"About your researcher's underwear," Canfield was saying, "I couldn't care less how much *creative writing* you do on your expense account. Just keep it out of my magazine . . ."

I can never really remember whether Rusk slammed the door behind him or not. Anyway, one second he was there, as big and fat-necked as life, and the next second Canfield said what he said and Rusk wasn't there anymore.

For a moment I was stupid enough to think that maybe Canfield had forgotten I was in the room—the way Rusk never even seemed to realize it—but I should have known better. Because after a bit he laughed and smiled at me and said, "Well . . . what do you think? Do you still want to work for this goddamn nut-house?" I knew then that that was one of the things about Jim Canfield. That when he was talking to you nobody else in the world existed. He talked to you as though he could hear what he was saying with your own ears—inside your own head. It was wild. It made you feel sad and beautiful and crazy, and you felt that somebody had given you the most tremendous and perilous present in the world. . . letting you meet this man.

"Magazines . . ." he said, then. "Nothing like them. Magazines have their own particular Cloud Nine. What's abnormal they consider normal; what's insane is for them perfect sanity. For instance, there was someone in here this morning—the ad director, as a matter of fact. He wanted to tell me he plans to release ten thousand balloons from the top of the Pan Am Building with the word 'Today' on them. Is that supposed to be sane? I mean if some kid in kindergarten thought it up it'd be a joke. And yet, the madness of it is that there *is* a reason."

He was quiet for a moment and then he picked up the paper Rusk had left behind and said, "How about that damnfool southern publisher of yours? Doesn't he beat everything, though?" It took a minute before I realized that what he was saying was that my boss, Tom Redford, must have been responsible for getting those advertising presidents to put in the ad.

"Tom Redford . . . ol' buddy Tom . . ." Canfield laughed as if over some old, terrible joke, but one he had gotten fond of in spite of himself. "Can't you see him conning those fat-cat agency presidents into each thinking it was his own idea? 'A little fittin' tribute to a fine ol' editor . . .' " His imitation of Redford's southern-soft rasp was perfect.

"Tom Redford," he said, "probably too stupid to realize how far out he's sticking his neck by indulging in just such a goddamn cavalier notion as that. Fool southerner probably doesn't have more than one or two threadbare principles to his whole name and then goes around squandering them on damnfool *gestures* that could just as soon get his own neck as fast as you can say Brad Rusk or any other suitable obscenity."

There was a copy of the latest issue of *Today* on the desk in front of him. "Crazy world—magazines," he said. "Like marriage in some ways . . . brings out the best in some, the worst in others.

"I've seen magazine people so dedicated to the stuff put down on that page, wild horses couldn't get them to change a word they believed in. You ever met Orin Kreedel—he's the executive editor here?"

I said no, because I had never spoken to him. But I knew him by sight.

"As surly as sin. If you try to buck him, that is," Canfield said. "But he knows what he's doing. We fight like born enemies at least once a week. But I wouldn't trade him for half ownership of the whole Kimberly shebang—not that that's worth anything much." He laughed a little grimly. "Orin Kreedel . . . he's been here with me for nineteen years . . . move into this office someday, I hope, when I get good and ready enough to move out—if ever."

His hands shook picking up the copy of *Today* and putting it down again. And I tried not to notice.

"On the other hand, there are always a certain number of the others . . . like our friend just left here now . . . the Rusks of the magazine world.

"Couldn't care less about the *idea* of a magazine. What does excite them are the 'goodies' that go along with the job—editorial privileges, to use their jargon. Doesn't matter whether they use it to get them into a ride at Disneyland without standing in line, or a plush boondoggle trip to Peru. That's what they eat up—that's what being an editor means to them. Free passes to shows . . . special service in restaurants . . . five-hundred-dollar teak doors . . . a secretary you can screw on the desk—like a pencil sharpener. Excuse me," he said, but not losing a beat, ". . . or letting it drop into the conversation that you'd skinned your knee on the side of the pool . . . the White House pool, that is, and isn't it a shame for the *image* that Johnson has so much flab . . . So help me, I heard one of them say that last week . . ."

He looked at me quietly then, dropping all the rest of what he had been talking about. "Forget about editors," he said, stopping to light another cigarette. Among the rest of the things on his desk there was a miniature gold tennis racket. For a moment, even as he stretched slowly to reach the cigarette lighter, I could imagine him hitting a crazy wild drive across the court in the blazing sun.

"What about you?" he said. "Something gives me the feeling

that you're always going to be up to your ears in *something.* Except why the devil isn't it your own life?"

I didn't answer him and he said, "You're a beautiful young woman. How come you don't have a man?"

I said, "I did."

"And . . . ?"

"I don't anymore."

He waited for me to go on and when I didn't he said, "You don't want to talk about it?"

"No."

"Well," he said, "that's up to you." And then, almost angrily, "Then why the devil isn't it your own writing you're involved with if that's what you know how to do?"

He stopped and waited for me to answer. Sometimes when people do that I can get out of it by just not saying anything. But not this time. He waited, and when I still hadn't answered after a minute he said, as direct as a knife, "Well, why isn't it?"

I stumbled a little and then started to say something about having to do something that would pay the rent . . .

"Do you *believe* that?"

All of a sudden I felt as though I were standing in back of an X-ray machine; even my teeth felt naked.

"Pardon me?"

"You want to pay the rent you get a job typing . . . or selling ribbons in the five-and-ten . . . or clerking in a bookstore. What you don't do is get yourself mixed up in a rat's nest like Kimberly. If you're a writer, you write."

"But, I do—I mean, at night——"

"When you're not either at Kimberly, or thinking about Kimberly, or planning about the next day at Kimberly . . ."

"But I've got to have a job."

"Not being some flunky managing editor, you don't. What did you think?" he said. "Did you think I've just been looking at your big brown eyes for the past fifteen minutes? I've been *listening* to you. Is that so rare in your life that you don't even recognize it? Don't give me any 'Kimberly's just a job for me' story. I *heard* you, remember?"

"I beg your pardon?"

"And stop treating me as though you were nine," he said. "You're at least twenty-five."

"Twenty-seven."

"So much the better. You can also stop looking at me as though you were some kind of a butter-wouldn't-melt-in-your-mouth little virgin. I'd make a pretty stiff bet that at heart you're at least a few percent bitch."

I don't know whether I flinched or not, but he said, "Mind you, there's nothing wrong with that. Nothing except, for some stupid reason, trying to make yourself think you're little Miss First Holy Communion. Come on, you're bigger than that. Otherwise I wouldn't be wasting my time talking to you."

I said to myself, "He's probably drunk . . . isn't that what they always say about him? Isn't that what everybody always says?" Except that I couldn't remember who, if anybody, I had ever heard say it.

"Look," he said, bending closer to me across the desk, "I'm an old man—and there's one luxury an old man can't afford. It's called kidding around. You're at Kimberly . . . and you stay at Kimberly . . . and you'll hold onto staying at Kimberly for just one reason. You know it. And I know it. So what the devil are we kidding each other about?"

I started to say something but there was no time. His words were like small, very straight, very sharp knives.

"You *love* it," he said. "The whole insane dizzy complication of it. Oh, maybe not the 'goodies'—with you I think it's something more subtle than that. You're a poor kid in a pretty damn sophisticated situation, and you want to see how far you can go with it. Well, great. Only stop lying to yourself—and trying to lie to me—saying you don't know how you got involved and you hate the idea of being involved. You *love* being involved in it."

I started to say something, but I never had a chance.

"And you'll never get uninvolved," he said. "Not until you've gotten it all out of your system. And for all you know, that may be never.

"One thing I do know," he said. "You'll never write again—nothing worth a damn anyway—until you go straight through this whole pit and come out the other side.

"Nothing wrong with being in the pit," he continued. "Matter of fact, it's a pretty good experience for a writer. Only one thing —you better pray to whatever God it is you believe in—and it might not be a bad idea to investigate *that* pretty well, too—you better pray you do come out on the other side. Either that, or spend your whole life being part of somebody else's phony involvement. I've seen it happen."

He was quiet for what seemed like a long time and then he said with absolute straightness, "If I'm wrong, why don't you tell me so?"

I still didn't answer him and he said, "Never mind. It'd probably be a lie—even though you'd probably believe it wasn't.

"As for now," he said, looking more *into* than at me, "you picked just about the most lunatic pit in the world to muck through. But there's one thing it can teach you—if you survive. And that is to see *people*. To come out from behind that phony fence you live back of. To see people—not just images or idiots or statues. People crawling and dreaming and scratching just to stay alive. And that means you, too. *And* me.

"It'll probably shake you up plenty, some of the muck you're going to get in the middle of. It'll take a peasant's ankle—and honesty—not to drown in it. But that's the only thing that'll give any purpose to staying here. You think you got it?"

It was strange how clearly every detail of that one meeting with Jim Canfield still stayed with me. Now, sitting back in the office, I knew that from the minute I'd walked in, there hadn't been any choice. That I'd have stayed at Kimberly no matter what the terms had been. Because—even though I disagreed with just about everything he said about me—I knew that I had never met anyone like that tigery old editor, and that in my whole life it might never happen again.

And all at once my skin went cold. He was dying. Suddenly I was as sure of that as of anything. That was why Rusk and Tom Redford and Wolffe were all here, closeted in that office down the hall. Brad Rusk's big moment of victory—and it was happening right now. Right this minute he was planning what he'd do with the magazine once Jim Canfield was out of the way. And in the whole wide world, what was there to stand between him and whatever he wanted to do? Jim Canfield? Except Canfield was dying—not dead but dying—not dead yet, because they'd never wait until the last minute to make their plans. Dave Wolffe? The unknown quantity in the whole situation? But Rusk seemed to have "savior David," as Canfield referred to him, tightly wrapped up in his own fat-hipped pocket. All that was left was Tom Redford—the biggest contradiction of all time. Redford . . . my honey-smooth, southern-pure boss . . . unpredictable . . . quixotic . . . as apt to blow violently one way as another.

I didn't like the odds. One unpredictable Mississippi gentleman against the infuriated ambition of a Brad Rusk coupled with the unquestioned power of his cocky unreadable friend.

Rusk didn't know anything about magazines; he had never even been on one before—in *any* capacity. His reputation—and there was no arguing that he had a "reputation"—had been gotten as a newspaper feature writer. For seven years before he had descended on Kimberly he had been a mainstay of one of the sleazier New York City dailies, writing articles that he had once described as "dedicated muckraking." His stock in trade was hate. And fear. Every series of articles he by-lined cashed in on some group hatred of some other group, and—with that as the foundation—he inevitably always had more than enough material to go around.

In the name of "honest, fearless journalism," he had done series of filth pieces on everything from, "The Avariciousness of the 'Seventh-Avenue-Sammy' " to "Does Your Doctor Prescribe Drugs with *You*—or the Graft-Heavy Drug Companies in Mind?" to "The Hidden Corruption on the *Other* Side of that Confession-Box Curtain."

The newspaper that printed the highly popular junk that vomited from his typewriter for seven years had finally folded just one month after the demise of *The Daily Mirror*, but not before Rusk had made a name for himself in journalism. Pretty much the same way you might say someone named Adolf Hitler had made a name in politics. So that it was with ultra-classic irony that when an ultra-conservative company like Kimberly had found itself in as desperate shape as it was, it had, with what amounted to criminal naïveté, turned to "the crusading editor"—"the fearless journalist"—as the instrument of its "salvation." Obeisant, full of last-ditch hope, Kimberly had given Rusk complete power with what amounted to a breathless, "Go forth, my son . . . and our blessings . . ."

And now, was Redford *all* that stood between Rusk's filthy kind of journalistic masturbation and the thirty-five years of work of that beautiful, tough old man, that old man they'd never have defeated in a million years—except in the one way? The odds made me sick to my stomach.

Except—was that really the *whole* team?

Like hell, I thought. There was still one more—one more that, God knows, they hadn't counted on any more than I had. One minor, unimportant, throw-away, screwed-up-but-fighting "appassionata" Italian whose mother had taught her better . . .

"Tough luck, New York Telephone Company," I said, tearing up the letter I had started. "I guess I'm just going to have to wait a little while longer before I can get around to enjoying those 'congenial surroundings.' " "*But not for the reason he said,*" I thought. And for a moment it was as though I could hear that old man laughing.

Having torn up my letter to the telephone company, I was finally able to get ready to go home. Redford's door was still closed and I stared at its ominous blankness all the while I waited for the creaking elevator to come.

The subway was crowded, and, half-hypnotized, I kept reading over and over again the Red Cross car card that I was facing. Help us Help, Help us Help, HelpusHelp, helpushelp, *helpushelp.*

As the train passed one of the local stops, my eye was caught by a poster advertising *Fiddler on the Roof*, and my heart tightened like a sponge being wrung out. I had seen it with Steve, while I was still seeing him, and now it was now, over, and they were still advertising the play. All that had happened had happened, and they were advertising the same play. I forced myself back to concentrating on the Red Cross card . . . helpushelp . . . helpus . . . helpus . . . *help* . . .

I walked down Eighty-eighth Street east to York Avenue, hardly seeing any of it. For all practical purposes I might still have been back at my desk.

When I went into the apartment my mother came and kissed me and said, "You work late?" I could smell linguini cooking in the kitchen.

My mother's name is Vivian, and she is Neapolitan. Today she had been dusting; I could tell by the white handkerchief she had tied around her head. My mother has a fierce, wild, Italian face; she also has very definite opinions about things.

"A little," I said, not knowing where to start.

"Better you should come home," she said, getting the dishes for dinner. "There's nothing for you *there*." Her hand made an eloquent gesture of disdain, dismissing all she meant by "there." "Better you get outside. Enjoy yourself."

I went to wash up and then came to the table in the kitchen. She had everything ready. I started to eat, and then I said, "The linguini's great," and concentrated on eating. Because I had made up my mind on the way home that I wasn't going to rehash the whole thing tonight about Kimberly with my mother. I had made that same decision several times before; so far I hadn't kept it once. Tonight, instead of talking, I'd get some writing done.

"What's the matter, nothing interesting happened at work today?"

I said, "Oh, I don't know . . . " and stopped.

She didn't say anything for about a minute, and then she got up to leave the table, and compulsively I said, "It really started this afternoon . . ."

My mother came back and sat down at the table and waited for me either to begin or not begin.

It was right then, sitting there, that I remembered how—until I was twenty-two—I had almost hated my mother's guts. Not that I had ever *thought* about it that much, it was just that, as soon as my father died, my mother suddenly went into this "respectability" thing. Not that she had had to push that hard to get me to stay home most of the time . . . usually I was just as happy reading a book as going to some droopy church dance. The only reason it had finally come to a head was because of Dorothy Johnson, one of the few girls in school I actually got friendly with. Just once I had brought her home with me. For some reason—maybe the fact that she laughed a lot, or had red hair, or who the hell knows why?—my mother had forbidden me to see her again.

It had been from that time on, until I was twenty-two, that everything my mother did to "keep me from making the wrong friends" . . . "to make sure I didn't get in with a bad crowd" became more and more exaggerated for me, so that as the years went by, and I stayed home more and more and read and wrote or whatever—even when there wasn't any choice, even when nobody else had particularly asked me to go anyplace—I had silently, more and more strongly, built up this hate against my mother and whatever the absolutely psychotic "respectability" thing was that she had. Until that day, when I was twenty-two, and I had come right home after work, and *he* had been there.

Now, I suddenly came back to where I was—sitting at the kitchen table opposite my mother, who was never *pushing* for information, because she never did that anymore—never, really, since the time after he came—just sitting, waiting, for me to begin, or not to begin.

"It really started this afternoon," I repeated. "I was just writing a letter to Jim Canfield . . ."

"Which one is he?" my mother asked quietly, fascinated by her own personal soap opera, bending across the table to hear.

"The editor of *Today*, you know . . ."

"Oh, the old one," she said. "Yes, I remember."

"Well," I said, putting down my knife, "I heard that he'd been taken to the hospital, and I was trying to write him a note this afternoon when Tom Redford came into my office . . ."

"The charm boy," she said. She had her own names for the whole cast of characters.

I didn't stop; in fact, there was no stopping me now. Slowly, eloquently, dramatically, I started to string out the whole story of the day, and fascinated, enthralled, and repelled at the same time, my mother sat absolutely still in the kitchen chair grasping and weighing every word. So that the whole house was completely quiet, except for me telling my story.

And, for the old man I imagined I could hear, somewhere behind me, laughing.

That night, lying in bed, I thought about that time when I was twenty-two, when I came home from work and from then on stopped hating my mother about wanting everything "respectable" after my father died. (My father—"*Daddy*"—the word mumbled into the darkness, the second's stab of pain as sharp as new loss . . .)

The man my mother was talking to was tall, and very wide in the shoulders, which were stretched over with a thick, high-necked sweater. His face looked like a grey skeleton with deep, burning eye-sockets. His chinos were rough, army-surplus, and next to him, leaning against the leg of the table, was a huge worn-out duffel bag. I just stood there staring at him, and after a minute my mother said, "Paula . . . this is your Uncle Sol . . . he's just in from the merchant marine . . ."

It was wild. Because my mother had never even said she had a brother before. The one relative she had talked about was a younger sister, Angela, who had died when she was twelve. My mother must have told me about Angela a dozen times, coming always very quietly to the place, at the end, where she had died of tuberculosis when she was twelve. There was even an old faded picture of Angela. My mother kept it between two pieces of glass on the bureau. But a *brother* . . .

After my mother had introduced us, a strange atmosphere had come into the room—something that I had not been able to figure out. They were both quiet, not a word, and finally it had dawned on me that maybe they couldn't talk because I was there, and I had taken a book and gone into the front room. After that the murmur of the two voices had gone back and forth . . . back and forth . . . and then suddenly their voices rose.

"The *filthy* habit," she had said. "I have no money for your filthy habit. I have no money even for myself. And even if I had it I wouldn't give it to you . . ."

And I had heard my uncle say, "What do you expect? Four years old I was when Papa died, and from that day on who ever even *saw* my mother? Work, work, work, and more work, that's all she ever did. Who ever knew that I even *had* a mother . . ."

And my mother saying, "So . . . so who took care of *me?* Even if there was no father . . . no mother that was ever home . . ."

A long silence. And then my uncle had said, "You . . . I guess you were different. You escaped. The only one. You found your own strength. But me—Angela and me—look how we turned out."

"You think I don't know what you're talking about," my mother had said. "You think when my husband died, I didn't see how easy the same thing could happen here." And somehow—without seeing—I had known it was my room she was gesturing to . . . that it was me she meant. "She thinks I don't want her to have friends—that I grudge her a good time—going out . . . What can I tell her? There's no father. And where that happens —where there is no father—I saw what can happen. I know—because I saw.

"What can I tell her? My own brother . . . on the streets . . . in jail . . . out of jail and then in jail again . . . taking dope . . . stealing . . . his arms like brown boards drilled with the filthy holes . . . Better never to talk about him. To think he's dead. Think you never had a brother, so she won't know. And Angie . . . Angela . . . the one she thinks died of T.B. when she was twelve. That she was twelve when she died—not seventeen. Because it is not that much a lie. Even when she was twelve . . . face like a cameo . . . cameo . . . Living on the streets . . . with a gang . . . anyplace . . . because my mother was too busy . . . working . . . working and working to make money to keep the family together. And when she was seventeen . . . from one of those places . . . from the roof . . . five stories . . . she threw herself down to the street . . .

"No—I don't tell her. Better if she hates me. Anything—anything—just so I keep it from happening again . . ."

In that front room, in that time, with the book in my hands, I had hummed to myself, I had made noises, I had screwed my fingers deep into my ears—because it was all there, going on in the kitchen, and I didn't want to know anything about it, I didn't even want to think about not knowing.

After that, how many hours? I didn't hear anything anymore. Until finally, my mother, almost yelling, "Go—go now—and don't come back. Never come back." And he had gone—stealing my class ring, out of whatever agonized need, from the top of the refrigerator, except I hadn't known until the next day—and after that I had gone right to bed, saying nothing, never going back into the kitchen that night.

After that it was as though my mother had *stopped.* Stopped telling me what to do . . . what not to do. Even when I went out and she didn't know where, she still *stopped.* All the questions there had been before. The rules. From that day—nothing. No questions. As though—that day, suddenly, hearing herself—she had recognized that keeping me from knowing almost anybody, doing anything—she had no more right. And the questions had stopped. And it had been from that day that I had started telling my mother things. About work. People I met. Details. Everything. Like—in a crazy way—she was making something up to me—and like—in a crazy way—I was telling her I understood. But never with words. Never—from either of us.

Lying there in the trackless connections of near-sleep, I heard Jim Canfield saying, "You're a beautiful young woman. How come you don't have a man?" And—still enmeshed with the thoughts of my mother—I remembered how even the two times Steve and I had gone away together for a week or so, she had never asked any questions about it. And suddenly, vulnerable to that attack when I had not expected it, I remembered the sharp, fragrant fragments of those times . . . Driving up together through New England . . . driving through that beautiful country . . . white-steepled churches . . . steaming mugs of coffee in small, warm-peopled diners . . . plaques in the square with the names of the honored dead. Beautiful, dream, red-leafed country of Woodstock, Burlington, Rutland, Bennington . . .

More than half asleep and yet restless, I thrashed in the sheet-twisted bed.

Once again, just before I went under, I thought I heard the old man somewhere behind me. He was still laughing.

two

I DON'T KNOW ABOUT ANYBODY ELSE, BUT one of the things I hate most in the world is waking up in bed alone. Even *before* Steve—even when there had never been anyone in particular to miss—for as long as I could remember, I had always hated waking up by myself. Anyway, it was a pretty close struggle the next morning to keep myself moving, with the rain Morse-coding against the window something that sounded like "go back to sleep . . . *go back to sleep* . . ."

My mother was getting up just as I finished making the supreme effort of getting into my dress, and she said, "Wait a couple of minutes. I'll make you breakfast." But I wasn't in the mood for conversation, so I said, "No, thanks. Anyway, they make better coffee than you do at Chock Full o' Nuts." She kissed me goodbye and I ran for the subway.

And so by eight-thirty I was in my usual red vinyl seat at Chock Full o' Nuts having my usual breakfast of coffee and one sugar doughnut (which, as far as I'm concerned, is one of the great tasting things of all time). I started to open *The New York Times,* but as usual I got hung up on those early-morning faces sitting around the formica-top islands.

This particular morning it was one of those construction workers with the overalls and the steel helmet that you see at noontime eating hero sandwiches and drinking half-pints of milk with their feet up on the sides of the construction, saying sweet hellos to the secretaries as they pass by, and making the pale, angle-pushing, expense-account lunchers seem kind of silly.

I was watching him break into his third cinnamon doughnut when Pete Larsen walked in.

Pete is about fifty-five, short and fat, with a surprised kind of eager-to-please, bunny-rabbit look and a great love for Tenafly in New Jersey, which, as he tells you, is " 'the Garden State' and the place where I was fortunate enough to have been born . . . except don't ask me exactly how many years ago, ha-ha."

Pete Larsen, head of the Promotion Department at *Today,* was technically the person I was responsible to. I often wondered how he, conservative to a fault, referred in his own mind to whatever he thought the relationship between Tom Redford and myself might be. He probably handled it in the "gentlemanly" way by juts not thinking about it—a typical old-regime Kimberly system of dealing with any and all "unpleasantnesses"—from affairs between publishers and staff members to a steadily growing yearly loss that was now reaching twenty million and gaining ground fast.

He smiled his usual little hesitant smile and came over, taking off his hat on the way.

" 'Morning, Paula," he said, placing himself gentleman-like on the stool next to mine.

"Good morning, Pete," I said. "How are you?"

He didn't answer, which seemed a little strange since the one thing you have to say for Pete is that he is polite to a fault. He ordered a cup of coffee and a brownie and stared down into the cup all the time he was drinking and swallowing, and even in between. His expression was even more strained than usual.

"Tom Redford said you were beginning to discuss where the

next sales meeting would be," I said, just to break the growing silence. "I hear he's pushing for that island—what's its name—Freeport—anyway, that place in the Bahamas—I think he likes the fact that you can gamble there."

"I was thinking of maybe suggesting Montauk," Pete finally said, blinking, "or one of those places in the Catskills."

"Oh." My mind drifted for a minute, hung up on the idea of Tom Redford in the Catskills . . .

He finished the last bite of brownie. "I have the impression Mr. Redford thinks I'm a little old-fashioned."

"Why . . . what do you mean?" I said, hating myself for the act I was putting on. With Larsen I always ended up getting angry with myself, and it usually started with my getting angry with him. Right then, as a matter of fact, it was almost a miracle that I didn't just come out with some stupid, smart-ass remark like, "I'll ask him about it the next time we're in bed."

Instead I figured I'd try it once more from the top, as if he'd just walked in and sat down next to me.

"I wonder if there's any word on how Jim Canfield is?"

For a second I didn't know if it was what I'd said or whether maybe he wasn't feeling well. All I got was one spooky expression, and then all of a sudden Pete Larsen was fumbling with his newspaper and losing control of it, and dropping it and ducking down onto the floor to get it back. Then he was pointing out to me on the messed-up, retrieved paper the place where it said: "Jim Canfield Dies. Editor of Today Magazine Succumbs to Heart Attack."

When I looked back at Pete Larsen he was still holding out the paper so I could read the headline, but he wasn't moving, and it was as if something had happened to his face, as if maybe it had started to melt. I tried to think of something to say, but each time I got scared stiff wondering what reaction I'd bring off. When I got to the third thing that I started to say—and didn't—and Pete Larsen was still just sitting there stock-still with his face melting, I did the one thing I'm frequently apt to do under trying and difficult circumstances when I'm scared stiff of what's going to happen next. I left.

I walked the two blocks to the office with my head down

against the wind that had started to blow up. Walking that way, I concentrated on each and every item in the street, using it to push away the one thing I didn't want to think about yet. Cigarette stubs—scraps of notes—torn receipts. Every litter bit helps . . .

Up in the self-service elevator, and down the hall into my office, I was still doing pretty well at thinking of things to keep myself from thinking; except that with usual Kimberly efficiency, the wastebaskets hadn't been emptied, and right on top of the mess in mine was my torn-up letter to the telephone company and half of an empty page that started, "Dear Mr. Canfield . . ."

I put my foot into the basket and squashed down the whole pile of junk. Stubborn as a mule, I swore to myself that I wouldn't cry, even as I felt the tears fuzzing my eyesight. With stiffly determined motions, I set about preparing my desk for the day—pencils, paper, carbon paper, calendar. As I ripped the top page off the calendar the first thing that hit me in the face was "November 12," and written right under that, "*Don't* call Steve." It was the same thing that was written on every page of the calendar through the end of the year. It had been scrawled at the top of each page for the past two months. Now—stunned—I felt for a moment the same way I had felt the morning I had written it over and over. Wild with anger and pain, all I had known was that I must never see Steve again—that after the argument of the night before there was nothing left between us except the need to hurt and be hurt. I had known without any doubt that if I did not call him he would not call me—not after the absoluteness of the way we had parted. At the time I had not even been able to imagine myself getting through to five o'clock without violating the law I had imprinted at the top of that calendar page—and now two months had passed. And I was right—he had never called. It seemed like a steep price to pay for being right. And yet it was not as though there had really been a choice. Not once it reached the stage where there was each time the constant, acid, eating-at-each-other corrosion because—loving him—there was always the never-go-away, not-to-be-lost fact that he was married—even worse, that he never once extended the temporary comfort of even lying about getting a divorce. That there seemed as little a chance of ever marrying him as there was of his ever actually going back to his wife, from whom he had been separated for seven years.

And those last martini-furious meetings in the same Third Avenue bar, Martell's ("visit Martell's-by-the-Sea"), the common denominator having now become anger—anger and argument—over everything, anything.

And that very last time . . . standing up among the drinkers at the long, lovely dark-wood bar . . . the open crocks of sharp cheese . . . the Indian boy bartender, wise, silent . . . Screaming suddenly. My voice . . . my words . . . something I could not remember. And the people who seemed not to hear . . . maneuvering past the narrow aisle next to the bar . . . the racks of foreign magazines . . . the dark door . . . into the street—my gloves left behind me, but I would never go back. And then the street. Windy . . . waving for a cab . . . "Hurry, let it come before he catches up with me—hurry up, for God's sake . . ." The tears hot in my eyes . . . the cold, face-freezing wind . . . blurring the street lights . . . and, from a block away, a cab catching my signal, turning to reach me, and then, sudden as sweat, "Not yet, *not yet* . . . give him *time* . . . dear *God* . . ."

And the cab stopping . . . and, behind me, the dark door to the saloon . . . dark . . . not opening . . .

The cab ride . . . cold, numbed, mumbling the address . . . the smell of cigar smoke . . . the ripped seat . . . the hack sign: Elmer Decker. Tacked to the back of his seat . . . printed in red . . . the letters sloping down . . . Please do not smoke. I am an asthmatic . . .

Now, restless, stunned by the memory of pain, I started taking out my aggressions on the typewriter, hitting the keys so hard that I cut the *o*'s right out of the paper. I slammed out almost two full pages of the photographic presentation I was working on, using all the *Today* promotion-type words that were already beginning to have a weird, familiar ring.

Promotion-type language for *Today* magazine included such words as *depth*, *graphics*, *perception*, *scope*, *far-reaching*, *penetrating*, *stimulating*. If you were talking about the people who read the magazine, or still better, its "audience," or even still better than that, its "demographics," you used words like *median*, *affluent*, *influential*, *decision-making*, *mobile*, *aware*, *urbane*, or—get this one—*movers and shakers*. Honest to God.

But even though the words filled up the page, my brain was still back there with Pete Larsen in the first minute when he held the newspaper out for me.

It was just about that time that Jack Sheehan arrived. Jack Sheehan, the promotion art director for *Today*, is probably only a few years older than I am. However, trying to be absolutely honest in the world of magazine promotion seems to have speeded up the aging process for him. It has also made him something of a drunk. I don't mean always. There were whole weeks at a time when he seemed to go without taking a drink. But one day, as the afternoon wore on and he still wasn't back from lunch, I'd remember that he hardly said a word all morning, just sat there at his drawing-board in the office we share, staring out of the window.

Invariably, around four-fifteen, he would return to the office. Smiling somewhat glassily, wearing his old, beat-up fedora at an angle that you'd try to convince yourself he has affected as a conscious joke, he would start talking about the South, a background factor he shares with our renowned publisher—after which all similarity ends.

He had returned that way from lunch yesterday. But I had been so preoccupied with Canfield's having been taken to the hospital that my usual fury at these four-fifteen drunken returns didn't quite catch on.

But this morning my anger at the sight of his flat-white, morning-after face felt like a stick of dynamite ready to go off in my head.

"Good morning," he said in that gracious, sheepish way of his, carefully putting the fedora away on the back of the shelf next to the filed Publishers Information Bureau reports and the half set of *Encyclopaedia Britannica* we had inherited with the job.

Deliberately being rude, I didn't answer him. He sat down quietly at his drawing-board, and I waited for him to say something else so I could have the satisfaction of being rude again. But he didn't, and that was the most maddening thing of all. He just sat perfectly still, his head bent down, and after a moment I realized that he was concentrating on Jim Canfield's obituary in the *Times.* With more of a need to scream than anything else, I said, as acidly as I could, "There's a theory that he might have *drunk* himself to death . . ."

The only thing worse than the way I felt before I said it was the way I felt afterward. Disgusted with myself, I shoved my chair away from the desk and plunged out into the corridor, heading nowhere, just "away . . ."

After about a mile and a half of inter-office walking, including elevator excursions and several side trips in and out of the various washrooms on different floors, I had worn off at least enough of the top layer of fury and restlessness to make me passable company for other people, if not myself. I was on my way back to my office when I met Jane Perry in the hallway. Jane was Tom Redford's secretary, one of those people who manage to be tall, blonde, intelligent, sexy, wise, and the epitome of a "lady"—even when she's using every drunken-sailor word in one sentence.

From the day I arrived Jane had, in her friendly but cool and ladylike way, befriended me—probably out of pity for anyone who seemed to be involved with Tom Redford. Of all the people I knew, Jane handled Redford better than anybody else. While maintaining all the distance and "respect" of a secretary, she nevertheless managed to get across to him the fact that he is not too bright, unread, Southern-bigoted, sexually loose but probably insecure, and somewhat lazy. All of this was conveyed in a calm, well-modulated manner I could no more learn to copy than fly.

"Hi," she said, "you look like a sunken navy . . ."

"Please, no poetry," I told her.

"Pretty rotten, isn't it?" She had a copy of the *Times* in her hand. It was a later edition than the one I had seen, and at the bottom of the obituary they had printed a formal epitaph. It said: "All the editors of Kimberly join me in expressing deepest sympathy at the passing of one of our colleagues." It was signed, "Brad Rusk, policy coordinator, Kimberly Publishing Company."

"Redford's pretty shaken up," she said. "He just called me from home."

"What's the matter," I asked her, "don't people die in the gentle South?" needing to hurt somebody, no matter what damn fool thing I had to say to do it.

You can understand what I mean about Jane Perry by the fact

that she didn't even bother to answer. Instead she said, "He's coming in about ten. There's going to be a meeting later this morning. I'm supposed to tell everybody on the sales staff."

"What kind of meeting?"

Very quietly she said, "It's to announce Mr. Canfield's successor."

"Who?" I could feel my *teeth* beginning to itch.

"Ted Monger."

"So that's what it was all about."

"What do you mean?"

"They were locked up together here last night when I left. It was pretty late."

"Redford and Wolffe?"

"And Rusk. Don't tell me he isn't going to be part of this celebration ceremony too?"

"Just Redford and Wolffe—as far as I know. And Monger, of course."

"The substitute shark."

"What do you mean?"

"Well, he's one of Rusk's boys, isn't he?"

"I don't know; is he? I don't really know anything about him. He works for *The Review*, doesn't he?"

"He's the features editor. Rusk hired him . . . another one of the 'young Turks' if I have my Kimberly pig Latin straight. I've seen him over there. Works like a steam engine. Lets Rusk make all the noise. Operates under the same flag as all of them—be a good little boy and someday Daddy'll make you an editor."

"Not about to give him the benefit of the doubt, are you," Jane said. It wasn't an accusation, or even a question; merely a statement. And one hundred percent true.

I went back into my office and sat down at my desk. Jack Sheehan was staring out the window. Five different times I tried opening a conversation so I could get around to making up for how nasty I'd been before, but I never quite made it. Every once in a while one of the salesmen would come into the office for one thing or another; half the time I figured they did it just to make brownie points in case the rumors about Redford and me were true. But sooner or later the conversation got around to Jim Canfield. The worst part of it was that they were all willing to have no opinion about Monger being made the new editor. I couldn't stand the wishy-washy way they all seemed to have agreed "to give him a chance."

I remembered now who it was that Monger reminded me of. It was a boy I had flipped about in my second-year chemistry class. Most of the time he never knew me for dust. The once he'd ever talked to me was when I passed him in the hallway and he said, "Hey, Jericho, your slip's hanging. What's it made out of anyway . . . an old Kleenex?" He and Monger would have gotten along just fine.

By the time eleven-fifteen came, I was just about ready to start a demonstration. We were all going into the conference room when one of the salesmen, Bill O'Neill, grinned at me and said, "You look mean enough to poison the milk." I turned on him, practically ready to start a fight then and there, but there's something about an Irishman stating the truth—and smiling—that makes you face the cool honesty of a fact.

I laughed back at him and said, "You better know it." O'Neill's remark had made me remember something else. It was a remark of Canfield's, part of which had gone, ". . . but one thing you'll never be, young lady, and that's a poker player."

For just that reason I decided to put as much space as possible between me and the visiting nobility. I squeezed into a seat at the very back of the room.

"Glory be, it's Judgment Day," someone said.

I didn't have to look to see who it was. His name was Rudolph Hecht, and he was in charge of public relations for *Today*. A tall, slightly stooped, balding man, he was a master at putting people down. One got the impression that he looked down not only on the magazine, but on the whole Kimberly company—including his own job—with a slightly amused disdain.

He made a point of cultivating only those individuals who could serve as a source of information for him. The rest he ignored with a kind of superior disgust. Some prime sources of information were the salesman who handled the liquor accounts, the ad director's secretary, and Helen Smith, the office manager, who could be relied upon to know *everything*, fiction as well as fact.

His way of cultivating these prime sources was to "trade" a

piece of information—or rumor—or possibility—for another piece of information—or rumor—or possibility.

Helen Smith had the office next to mine, so that frequently I could hear this exchange of valuable information going on. Given to affecting an English accent, Hecht, I guess, must have had a certain kind of elusive charm, since so many people seemed to think so, but it escaped me completely. As far as I was concerned all that he looked like was exactly what I had heard him reported as being: hypochondriacal, a nit-picker, sex-obsessed—probably impotent.

I had just about gotten seated when Redford and Wolffe and Monger arrived. They made a strange trio—Redford, outwardly, almost arrogantly calm; Dave Wolffe, tall, dark, unreadable, vaguely shady; and Ted Monger, red-haired, solidly built, attractive in a high-cheekboned, dark-skinned, Indian kind of way—Comanche, probably.

Right behind them came a small retinue of people, most of whom seemed like complete strangers until I realized that the tall, bald man in the center was Mr. Arthur Jefferson, chairman of the Board of Directors. I had never seen Jefferson before except in pictures, but he was obviously, unmistakably who he was. His head was as bald as a grape, with eyes as placid-blue as a one-year-old's. Seeing him, I immediately looked for the two people who—according to everything I had ever heard—were never more than ten feet away from him at any meeting. And, sure enough, just a little way to the side and in back of him was Mr. Wilfred Graves, a skeleton-thin man with a little black notebook which he grasped to his chest like a nun hugging her missal. Word was that the notebook contained every statistic relating to Kimberly for the past fifty years. He had connected with Jefferson, or so I had heard, when they were both accountants in an insurance firm.

The other person I looked for—except that you didn't have to, she was right there—was Miss Lorraine Clapper (which she pronounced Clay-per). Secretary to Jefferson, she had, I'd been told, monumental influence with him. She sat down directly next to him now, a mink cape ("What a cop-out," I thought, "a *cape!*") hugged tight around her shoulders even though the room was stuffy and hot. She was twenty-two, blonde, and for six months just before "connecting" with Jefferson had been with a road company of *Carousel.*

Redford started the meeting. It was amazing how carefully you had to listen to what he said to realize how stupid he was.

Today, wisely for once, he played it very quiet and cool. It was wild, since I knew how little actually got *through* to him, but he seemed honestly shaken by Jim Canfield's death. I remembered the sharp but undeniably fond sarcasm with which Canfield had spoken of Redford; what would seem an impossible bond had been, in some crazy way, established between these two opposite poles.

"You'll all be happy to know that today we're being privileged to have with us Mr. Arthur Jefferson, our very fine chairman of the Board of Directors of Kimberly, and his assistant, Mr. Graves," Redford said in opening. "We're also being graced by the fine presence of Miss Lorraine Clapper. I'd like to express to them how happy and pleased we are to have them here."

He tossed back the lock of blond hair that was kind of an unconscious prop and said, "I think you've all met Mr. Wolffe, the chairman of the magazine division." Wolffe nodded noncommittally. No one said anything, and after a moment Redford continued. "Mr. Wolffe came over himself this afternoon so he could personally introduce you to the new editor of *Today.* Some of you probably know Ted Monger here—he's one of the geniuses responsible for putting together *The Review* every week. ("Watch the death-wish, Redford," I muttered.) He's been doing edit work for a good many years at Kimberly and elsewhere, and I'm satisfied that he's going to do a fine job for *Today.*" He stepped aside. "Well . . . I'll let the gentleman speak for himself."

Monger didn't say anything for a moment. It was a sticky position for anybody to be in. Canfield, as they say in show biz, wasn't going to be exactly an easy act to follow . . .

"O.K., Monger," I said to myself, "you're on. Let's see you go." Except that before I could even get my teeth set in a nice, healthy meanness, Monger had said, very quietly, "I don't know anything about *Today,* but starting right now I'll try to learn. I promise each of you here that I'll be the best editor I can. Since it would be stupid to talk about a magazine I know nothing about, I won't." Then he picked up his briefcase and walked out of the room.

Filled with the frustration of having been gypped out of my most obvious villain, I pounced on the next most likely object for my bursting-at-the-seams fury. And Dave Wolffe seemed to have been born for the part.

I studied him like someone taking inventory, the general idea being to come up with as many nasty points as possible. It wasn't too hard to do. Tall, straight, with dry-looking brown hair . . . cold, I-don't-trust-anybody eyes . . . *carefully* dressed . . . small, stubby hands . . . smooth, taut skin that looked as if it would be cold to the touch. Everything added up to someone I didn't really have to *work* at disliking.

"I know there have been rumors," he said, "so the first thing I would like to tell you is that—no matter what you may have heard—I have not been hired to liquidate Kimberly." Damn, I thought; it wasn't going right. But there was still time . . . "I never really had the privilege of knowing Jim Canfield," he said. "I envy those of you who did." God damn the con man, I thought; but I was listening . . . closely, carefully, as intently as any of the rest of them in the room, I was *listening*. "Please," I prayed, "let him do something *wrong* . . ."

Instead he went on, as simply and quietly as he had started. "Obviously I know *Today* less well than any of you, so it would be absolutely presumptuous for me to say anything about the magazine. What I would like to say is that I firmly believe that Ted Monger will do everything he can to be the best editor he can. And that—to whatever degree I can speak for the attitude of the management—*Today* will have its complete and absolute support—and no strings attached." Then he simply stopped talking and sat down.

A second later Redford stood up. He said, "I want to thank both of you gentlemen for coming and talking to us today. I'm sure if there was any of us had any reservations about how things are going to be handled, both of you have put those fears to rest."

He turned to his right. "Now, Mr. Jefferson, I wonder if we could be so bold as to ask for a few very valuable words from yourself . . ."

Jefferson cleared his throat and then, speaking right from where he sat, said, "All I'm interested in is the figures. I've had a saying all my life, and it fits in here, too. It goes: You don't have to love us, just pay off. The same goes for *Today*. All I'm concerned with is the color ink we have to use in the ledgers at the end of the month. That's all I got to say."

No one said anything. I saw Miss Lorraine Clapper reach out and—so help me—brush some lint off the back of his collar. As she was doing it she smiled. I'd never seen a smile exactly like that before. It was sort of like the Mona Lisa, assuming that what she's got hiding in that picture happens to be a little pearl-handled revolver.

The brass, headed by Mr. Jefferson and Miss Clapper, left right after the meeting, and we all filtered back to our desks in a vague sort of quiet way, nobody saying anything very much. Feeling vaguely uncomfortable and dissatisfied, I went back to the presentation I was working on; even at that the words had a way of all seeming to run together. "It is a fact that *Today's* affluent educatedmotivatedaudience . . ."

I had been jabbing insistently at the typewriter for the better part of an hour when I looked up and saw that Redford had come into the room. I glanced over at Jack Sheehan's desk, thinking he had been there all the while, surprised now to find that he wasn't.

"Well," Redford said, settling himself into the chair next to my desk, his elegant, basset-hound face still much more attractive to me than I liked to admit. "You sure costin' us a lot of wear and tear on that machine. You think you got any idea what you're doin'?"

"I hope so," I said, the uncharacteristic blandness of it surprising me probably more than it did him.

Seeing that I wasn't going to rise to just *any* bait, as was almost always predictable, Redford recrossed his long, southern-skinny legs in a typical "lord of the plantation" manner and started off fresh.

"Well, Miss Paula, how did you think I did in there?"

For a second I was completely sure that he was putting me on. I took my time looking up from the typewriter in front me, giving myself a breathing space. Then I looked right at him, decided he wasn't just itching for a fight and said, "I thought you were fine."

Neither of us said anything for a minute then. It was just a matter of waiting long enough—except that for the first time in my life I was the one capable of waiting the one second longer. I was just taking a breath to say God-knows-what when Redford beat me to it.

"For somebody claims to be a writer, you sure have kept yourself a damn closed mind."

"What about?"

"What do you mean what about? What gives you the right to be so Goddamn sure that's not going to be a mighty good editor we heard in there this morning? What *facts* you got to go by?"

"You mean just because Ted Monger *happens* to be Brad Rusk's boy?" I said. "Just because putting him into the job will be exactly the same for Rusk as if he were running *Today* himself?" It felt good to be saying all these things; once again I knew where I stood.

"Listen," Redford was saying, "you listen here to me—" and the funniest part wasn't that he was having this kind of unmatched fight with me. (In this corner, Tom Redford, publisher at ten million dollars a year—and over here one smart-alecky promotion writer, recently fired and hired and who knows what next . . .) All *that* had happened before. What was funny was that he wasn't really fighting . . . he was talking quietly, even persuasively; there wasn't a single sign of a roar. It was as if he were truly and quietly convinced of the truth of what he was saying.

"Ted Monger's a good man," he said. "I wouldn't have agreed to his being editor if I wasn't convinced of that. I don't know what conclusion you've come to in your own mind, and I couldn't care less, but Ted Monger's editor of *Today* because I agreed he could be. And the only reason I agreed he could be is because I think he's going to be right for the job. You hear me?"

And I thought: But what about Orin Kreedel? What had Canfield said? "As surly as sin if you try to buck him—but he knows what he's doing . . . Move into this office someday . . . when I get good and ready . . . *when I get good and ready.*"

What about *him*—what about Orin Kreedel? But Redford's voice was going on—he was standing up then, short, straight, his blue eyes like marbles with sun on them, and as he left he said, "You want to do something for Jim Canfield? Instead of fighting them, why don't you stay around and help them do what he started?"

It sounded good—except I wasn't sure. Not yet. Not quite.

But what if he was right? He *could* be. God knows, nothing that had happened at the meeting this afternoon had proved him wrong. But the best I could come up with was just knowing that I didn't know.

Ten minutes later, still as confused as before, I left my desk to go to the ladies' room. As I went in I heard a sort of moaning sound.

It was Jim Canfield's secretary, Marie Davis. Her face was almost brick-red with crying—her eyes looked all torn apart, and she was shaking so much you could barely make out what she was saying. "Ted Monger," she said . . . "Ted Monger." I don't think she even knew who I was—or who she was talking to—she just kept saying it over and over again.

"Monger—Ted Monger—he said get all that old man's—get all that old man's junk out of the office by five o'clock. That's what he said . . . throw it out . . . get rid of it . . . *get it out* . . ."

I slammed myself into the nearest john and puked.

three

ONCE, ON THE UPTOWN IRT, I HEARD AN old black woman say—to no one in particular—"The way I feel right now, was I to start cryin' I wouldn't stop till next Fourth of July . . ."

Those were pretty much my sentiments after I heard what I'd heard in the ladies' room. I could almost *feel* the way Monger must have yelled. "Get that old man's junk out of here."

I went back toward my office. Sheehan and most of the others had left for lunch. For at least five minutes I just sat there at my desk in the empty office feeling the anger and hurt and lostness washing over me.

Out of desperation, I decided that I'd call up somebody to have lunch with. The first person I thought of was Louise. I called her office and she wasn't there, so I called her at home.

Louise and I met in high school. She's about my age, red-headed, and full of the honest-to-God belief that "everything will be all right." She actually says so in so many words. Sometimes I come pretty close to hating Louise, not just for believing *that*, but for having a husband like Bill, plus two completely unobnoxious kids. I guess if we didn't fight all the time we might not even be friends. As it is, we usually yell at each other and everything ends up O.K. After Steve and I broke off she never once said, "everything's going to be all right" during those first few weeks when I couldn't even breathe without it hurting. It must have cost her a lot doing that.

"What's the matter with you?" I said now over the phone.

"I've got a cold . . . except I think it's pneumonia," she rasped.

"Oh . . ." I said, feeling suddenly lost—way out of proportion.

"What's the matter?" she said. "Paula? You sound funny."

"Nothing," I said. "I mean—I phoned your office because I thought maybe we could have lunch."

"Gee, I'm sorry, Paula. I'll probably be in bed with this thing for at least a couple of days, whatever it is. I'd ask you to come here only—well, I feel as contagious as if I had nits. I don't think you should come up."

"Well . . . no . . . I guess not," I said. "Take care of yourself."

"You, too."

"Well—so long."

"So long, Paula." She was sneezing again as she hung up the phone.

I couldn't think of anyone at Kimberly to call, so I pulled out my little red telephone book and started going through it right from the beginning.

Finally under the *H*'s there was the name of a television producer I'd met once at my agent's Christmas party. We'd gone out together a few times, and he might not even remember who I was, but I decided I'd try it anyway. I dialed the number which started out CI6 . . . While I was doing it I made up a little cheery, throw-away opening that went, "Hi . . . guess you didn't know this was going to be your lucky day." On the fourth ring a very musky-lady type voice said sleepily, "Hello? who is it?" and I hung up the phone.

Under the *N*'s there was the number of a drugstore that was open on Sunday and under the *R*'s was *Ricci.* Ricci was the pizza place in my neighborhood that delivered. And that was it.

I sat there for a solid two minutes without moving a muscle.

Then, without even thinking, I picked up the phone and did something I only do when I really find myself drowning. For alcoholics, I guess, the same thing would be calling AA. For me it takes the form of dialing CI6-4200.

The recorded voice said: Thank you for calling the Fifth Avenue Presbyterian Church. Let us pray together . . .

The record was thin and high-pitched. For a minute I tried to imagine the face of the person speaking, but nothing came up, and by that time the record was over, and all I'd heard was ". . . and God give me courage to face this day. Amen."

At least that was *something.* I still didn't have anything special in mind, but anything was better than staying around that empty office, so I grabbed my coat and bag and started toward the elevator.

I was trying hard to make myself think of something besides Kimberly when I got downstairs and saw Rusk and Monger standing in front of the directory in the lobby. One of the building maintenance men had taken the glass front off the directory and was rearranging the names. I tried walking past without paying any attention, and I'd just about managed it when I heard Rusk's phony-gutsy voice saying to the maintenance man, "Never mind saving that Canfield one. We won't be using that again."

I practically ran, but even so I couldn't miss Rusk's filthy laugh following me down the lobby and out into the street. And all of a sudden I found myself drenched in the sickening memory of the first time I'd ever met Brad Rusk.

It was when he'd just been made policy coordinator of Kimberly and while I was still managing editor on *Woman's World.* It wasn't more than two days after his appointment that Rusk called a meeting of three or four people from each of the different magazines. When we got to his newly carpeted, cork-walled, teak-desked office (Fritz Mott and the art director and I were the three representing *Woman's World*), Rusk was standing in the middle of the room, his feet about a yard apart, his chin stuck out, his hands behind his back like some kind of damn sailor or something, not saying a word until all of us sat down. And not for at least three solid minutes after that. It was during that time that I noticed there wasn't a single person in the room from *Today*.

The first thing Rusk finally did was to pick up a magic marker—it was a green one—and walk to his desk and look down at a yellow pad that was lying right near the edge. "Point number one . . . *Art!*" he announced. After which, taking three steps over to a big drawing pad that had been set up on an easel in the middle of the room, he spelled it out laboriously, printing it in letters the size some two-year-old would make, "A-R-T . . . *art* . . ."

He'd gone through exactly the same ritual fourteen times, "Point number two, *Photography*" . . . walk to the easel . . . printing photography . . . "P-H-O-T-O-G-R-A-P-H-Y . . ." back to the desk . . . looking down at the yellow pad . . . "Point number seven . . . *Mail!*" Back to the easel . . . "mail . . . M-A-I-L . . ." On and on, creaking the soles of his huge black suede shoes, scuffing the new carpeting.

When he finished, he stopped, scratched his behind, and said, "Now, for instance . . . let's take point number five . . ."

I'd looked over to the easel to see what point number five was. It said: "#5. FICTION." I'd looked back to Rusk. He'd hesitated, heavily silent, giving everyone a chance to let Number Five sink in.

"Now," he said at last, "fiction." I'd gotten so used to the routine I was almost startled not to have him spell it out again. "At the moment," he said, "we're using fiction in *The Review* and *Ladies Fair* . . . and it isn't impossible that sometime soon we may be including fiction in *Today*."

Nobody except me seemed to find anything odd about that either; at least not so you could tell it from their faces. I was in the middle of considering the strangeness of Rusk's statement when the door to Brad Rusk's office opened and the esteemed president of the company (and Tom Redford's ol' buddy), Bob Moshier, stuck his head into the room. For a moment I saw Rusk stiffen, stop in mid-sentence, and then finally say, "Yes, Bob? did you want something?"

Moshier didn't say anything for a split second, then immediately he grinned, his face all salesman charm, and he sort of ducked and said, " 'Scuse me, guess I'm in the wrong place."

"We're holding an editorial meeting," Rusk said, sharp, snotty, not giving an inch—just standing there, waiting for Moshier to back out again. Which was exactly what he did.

Fumbling just a little, Moshier said, "Which I will then leave in your very capable hands, Brad . . ." and went, closing the door behind him, his big Dutch-Boy-Paint grin seeming—like the Cheshire Cat's—to remain in the room for at least ten seconds after he had left.

The whole thing had taken only a few minutes, but for some reason I had the feeling that it was indicative of something much more important than it seemed to be. And for just a second I wondered whether it was going to be possible for two such egomaniacs as Rusk and Moshier to work together and keep out of each other's hair.

Immediately after the interruption by Moshier, Rusk had continued with his little speech.

"Now it's stupid," he said, scratching with his left hand between his shoulder blades, "it's stupid to have each of the fiction editors of all those magazines purchasing fiction *separately*. Christ, we might even be bidding against ourselves without knowing it. My way," he said, "the *Rusk* way . . . the only logical way . . . is this. Take the William Morris agency, for example. Instead of three editors buying fiction separately, we'll have say one man . . . one man who goes in to William Morris every week and simply says, 'What fiction have you got for Kimberly?' " I remembered not believing my ears. I was sure there was a joke in it someplace.

Except nobody was laughing.

I looked at the art director. He was staring down at the carpeting. I looked at the editor, the man I worked for, Fritz Mott. He was sucking furiously on a short, thick cigar, even though it was obvious it had gone out a long time ago. I got the impression he was wishing his wife, Maud, were there. Lots of times he had this thing about just sucking on a cigar instead of saying anything. Once I had heard somebody say that was how he had gotten his job in the first place.

"Incidentally," Rusk was saying, with a great, dramatic pause,

"I think it would be wise for you to keep in mind that I'm not merely speaking academically. Just in case any of you here have any doubts, let me present my credentials." For a moment I thought that he was going to give a whole presentation on Brad Rusk. I wasn't too far off.

"When I was growing up," he said, "my father had a small candy business in Williamsport, Pennsylvania. And a good reason why the business *remained* small was my father." The way he said "father" made you understand what the word *disdain* really means. "The whole thing was straight cliché," he went on mockingly. " 'Secret' family recipes . . . high prices . . . exclusive rich-ass clientele. 'Hill' people we called them. Not that they ate all that much chocolate, but when they did, it had to be the best. And that's what my father fed his starving, puny little ego on. And God knows that's all he fed. We were so poor even the rats couldn't make shit out of what we left.

"The year before my mother died we were half an inch away from starvation. That was when I offered to take over the business. I was fourteen." He stopped for a full minute, his thick greying eyebrows close to meeting in a solid slash across his forehead.

"My mother thought it was a joke . . . she called me a *traumerisch* . . ." The bitterness in his voice now was like a razor-sharpened pick striking into ice.

"In a year it had all changed. Prices down, more people buying (and we didn't particularly care *where* they came from), and no more precious, secret, *expensive* recipes. A single year . . . Sales were up, there was meat on the table, and—most important of all—we could afford to tell all those precious, prick *patrons* exactly where they could take their patronage and shove it.

"Of course, she was dead by then—but she *knew* . . ."

I felt the hairs stand up on my neck. But then, abruptly, like someone coming in from outside and slamming the door behind him, he was back in the room. Brash, blatant . . . inescapable—*evil* was the only word that really fit.

"And that's what I intend to bring here . . . to this weak, puny, rotting organization. Consolidation. Efficiency. Figures in the black.

"I'll show her." And there was no need to ask who he meant.

Now, having gotten his second wind, he was further explaining how his theories would work specifically; how it would only make sense that if *The Review*, for instance, sent a photographer to Spain to shoot the street riots, he could just as well get pictures of the Paris showings for *Ladies Fair* on the same trip. It was like saying to Durante, "As long as you've got your mouth open you might as well sing 'Traviata.' "

I remember thinking, "I probably just heard wrong," when I realized I'd raised my hand.

"Yes??" Rusk said.

"Mr. Rusk," I said, "I guess maybe I don't understand exactly what you said, because . . . well, the way I thought you said it, after a while you'd be making all the magazines like all the other magazines, wouldn't you?" It occurred to me while I was saying it that the grammar was a little weird, but all in all it was what I meant to say.

I could still remember how quiet it had gotten then. Nobody said a word. Rusk didn't move at all for a moment—then he raised his hand and ran his chubby fingers through his straight black hair. Finally the silence was broken by Rusk laughing—or at least something that was supposed to be a laugh, even if it did sound a little high and hysterical.

"You mean," he said, stumbling over the phony laughter he'd manufactured for himself and then starting again, "you mean point *seven?* You mean it might make the magazines all seem alike because we're going to use the same *mail room* instead of four different ones? Is that what you're worried about, Miss?" and he laughed—or whatever it was—again.

I remembered that I'd started to say, "I wasn't talking about the mail room. What I meant was everybody using the same writers and photographers and . . ." Right now I couldn't remember exactly what had happened at that point except that all of a sudden the meeting had ended very abruptly. Rusk and I had for the first time, as they say, "interacted."

Now, forcing myself to swallow deeply, I made a wild effort to forget all that. I stood absolutely still in the street, which was full of people out on their lunch hours, counted to three, and started to walk very quickly *away*.

The Avenue was crowded the way it was every lunchtime. All of a sudden, passing a girl and fellow holding hands, I felt like I was drowning in loneliness . . . as though they had stolen something from me. Right then I decided to do something about it. I'd chuck the whole sick pattern. I'd be exactly what I'd taken the job at Kimberly to be . . . a real writer (that meant words down on paper, not *thinking* about it) who also made $125 a week writing promotion copy for *Today* magazine. Period. (*Incidentally, I'd show Jim Canfield . . .*)

I picked up some cottage cheese at the deli across from work on my way back, ignoring the long, steamy counter where the fast, hairy-armed men slapped together their sandwiches of hot pastrami and liverwurst and roast beef. There still was almost nobody back at work, and Jack Sheehan's desk looked even emptier than it usually did. After I'd finished eating my cottage cheese I found myself thinking, the hell with Kimberly . . . I had my own life to lead. I'd make some friends I could have lunch with without its being a major production. I'd start reading some more . . . reading and *writing.* All of a sudden I got furious thinking of the seven years I'd wasted.

Paper—the first thing I'd do would be to take a whole stack of paper home from the stockroom. Paper, and carbon, and clips, and—I stood up, pushed back the chair, flung away the empty cottage-cheese container, squared my shoulders, and marched straight down the hall toward the stockroom.

Five minutes later, laden down with at least one of everything I could find, including three different weights of white paper and a bottle of something called snopake Correction Fluid, I got back to my office, planning how I'd leave at the dot of five . . . it would only take me——

"What you doin' honey? Figurin' to start a little magazine of your own, are you?"

I almost dropped everything I was carrying, but all I actually lost hold of was the red-and-black Royal ribbon.

Tom Redford picked it up, looked at the box, and then placed it carefully on top of all the rest of the loot I was carrying. "You forgot erasers," he said. "Or are you jest planning on letting it stand as it comes?"

I didn't answer him. But rather than taking insult at my rudeness he eased himself into the big brown leather chair next to my desk.

For a moment I thought about arranging the paper in my desk drawer; at least that would give me something to do. But I decided the whole thing was just too damn dopey, to say nothing of phony, and so I started packing all the stuff into a brown paper envelope instead. Once while I was doing it I even looked up at him; it was as if I couldn't keep myself from daring him to say something. But he never mentioned it again. Instead he just sat there smiling. "The hell with him," I thought, "the hell with all of them." Right after work I'd go right home and start writing. I might even do a whole chapter tonight. All I'd have to do would be to read over what I'd done so far and then—I might even——

"You ever goin' to answer me?" he said.

"I beg your pardon?" I looked up at him; he was smiling in a smug kind of a way. "I'm sorry. Did you say something?"

"I jest been sittin' here inviting you to take a drink with me this evening after work," he said. "You think you could manage to do that?"

"No," I said, "I mean—well—" (*I couldn't remember the name of the woman in the book . . .*)

"Well—yes or no?"

"No," I said. (*Fortuna . . . was it Fortuna? . . .*) "I mean—well—I don't know—I mean—yes. Yes—thank you."

"Good," he said, standing up, grinning, stretching expansively. He started out of the room. At the doorway he stopped. Then he came back and picked up the two typewriter ribbons still sitting on my desk and placed them very carefully on top of the brown envelope I'd stuffed full of paper.

"Don't forget these," he said. "'Less you goin' in for ghost-writin'." And smiling, southern-style, he left the room.

four

SNOW STARTED TO FALL ABOUT TWO THAT afternoon, and by four-thirty the place was emptying out fast. Redford had called me in to his office. Right after I sat down he had handed me a piece of white paper. On it was written in impeccable tiny print: "Two ormolu candlesticks, circa 1800, $12,000. One Ming vase, miniature butterfly design, $5,000. One Meissen soap dish and brush-holder, $8,500." On the bottom was written: "The Max Spillman Gallery, 344 Madison Avenue."

"I tell you what I want you to do," Redford said after I had read what was written on the paper and registered complete confusion. "Sometime tomorrow I want you to go to Mr. Spillman's gallery and look at those three little geegaws. You got pretty classy taste. I want you to pick out one of them goodies and have Mr. Spillman send it on to Miss Ginny Eisman at that advertising agency she works at. It's in exchange for a little favor she did for me.

"Miss Ginny's a pretty good writer, wouldn't you say?" he asked me, a sly kind of look around his eyes.

All I knew about Miss Ginny Eisman was that she was always having her picture in the Sunday magazine section with a cigarette holder and raised eyebrows and a long kind of I'm-too-English-to-care-about-money face, over a caption that usually started, "Ginny Eisman, vivacious young vice-president at the Charles and Mark Advertising agency . . ."

I didn't know she was a writer," I said.

"Fickle hussy. Fact of the matter is, I happen to know you're a real ad-mirer of her work." He opened the top drawer of his desk and took out a folded sheet of newspaper and handed it to me.

I unfolded the paper, feeling more than ever like Alice in Wonderland, positive that what would be printed there in twenty-four-point Old English caps would be "EAT ME." Instead it started, "A Salute to Jim Canfield on Thirty-Five Years of Today. We, the undersigned members of the advertising fraternity . . ." It was the newspaper ad that Brad Rusk had practically gotten apoplexy over that day in Canfield's office.

I said, "You mean—Ginny Eisman wrote that for you?"

"Let's just say I . . . *inspired* her to write it." He laughed out loud. It was like a sound montage of southern insolence, dirty humor, and just plain meanness.

"Anyway," he said, "what I want you to do tomorrow is arrange a little thank-you present for our friend, Miss Ginny. Maybe the two candlesticks would be best. Then she'd be twice as grateful . . ."

"But—they're twelve thousand dollars."

"We just take that out of old Pete Larsen's promotion budget, honey—that's what it's *for*. And here I was, thinking I was teaching you not to think poor anymore. Never mind," he said, putting his arm around my shoulder, "we gonna get rid of them poor white trash ideas of yours yet. Not your fault. Probably picked them up working with them tacky edit people. Come on now," he said, "let's us get us that drink we been talking about."

I got my coat and things and when I got back Redford was ready and waiting for me. I thought everybody had gone home, but when we passed the office of the ad director, Mark Post, he was still at his desk.

Post was a slight, blond-haired, rather good-looking man, but for some reason I could never accept him one hundred percent. I thought maybe it was that he always looked wide-eyed confused when he saw Redford and me talking to each other.

" 'Night, Mark," Redford said as we walked past, and Post looked up, peered at us for a moment over his black-rimmed glasses, as though he weren't quite sure who it was, and then said, "Well—going home, Tom? . . . Paula?"

It seemed like a pretty stupid question to me, but Tom just laughed and said, "No reason us hanging around; you doing all the work anyway," and kept on walking. "Too busy even to care whether he gets laid or not . . ." he commented in a sort of half-whisper.

At the elevator, almost as though he sensed I might take what he had said as being less than "respectful"—a word that has a Southern connotation that you either have to be born there to understand, or forget it—he said, "Now there's a man could be a writer, publisher, editor—anything he wanted. One of the most intelligent men you ever going to meet." When I didn't disagree with him, or even answer, he added half under his breath, "Only trouble with Mark Post is he's never been to the fight."

It was the kind of Tom Redford statement that didn't mean very much at the time but that I was to remember a long time afterward. All I knew then was that Mark Post had been manager of the Atlanta branch office almost since *Today* had started, that he was a close personal friend of Jim Canfield's, and that if Bob Moshier, the president of the company, hadn't been an old friend of Redford's—a "*bud*-dy"—Mark Post would probably be publisher.

But, for the moment, the subject I couldn't get out of my head was the one we had been discussing before—the one that was to precipitate my little twelve-thousand-dollar purchase the next day. *Surprise, surprise, Pete Larsen . . .*

Thinking of it, I said, "By the way, Pete Larsen was saying he felt you thought he was old-fashioned. He seemed worried about it."

"What ever gave him that idea?" Redford said, his voice pitched to the height of bewilderment. And then, one second later, "It's the honest-to-God truth—but whatever gave him the *idea?*"

I took a step backward and said, "Sometimes I get the impression there aren't very many people you *do* like."

He turned sideways in the elevator so that he was facing me head on.

"Sure, I like people," he said. "Just so long as they're Baptists, white, and come from Mississippi." He said it so straight-faced I didn't even laugh.

We had reached the lobby now, and I could see the snow still falling heavily as though it would last through the night.

"I have just one question I'd like to ask," I said.

"What's that?"

"How'd you get a list like that?"

He hesitated, as though maybe his first impulse was to go into his "you don't know your place" routine, but then he took a deep breath and said, "I'll tell you how it is— Every man—or *woman* —is corrupt. It's just a matter of what they're corrupt *about.* With some men it's money. With some women it's men. And with some it's china . . . Now," he said, dropping the subject once and for all, "where you gonna buy me that drink? The Rainbow Room tidy enough for you?"

I agreed that it was, and we walked the rest of the way there through the snow without saying a word.

As soon as we walked into that wild room with all those windows and practically miles of smoky gold mirror, Redford grandly accosted the nearest waiter.

"You got a table with a long tablecloth?"

"I beg your pardon, sir?"

"I said have you got a table with a long tablecloth for my fiancée here and me?"

Without saying a word, the waiter led us to our seats, at which point Redford promptly excused himself, announcing as he left, "Now don't go pickin' up any sailors while I'm gone. Remember what happened last time."

I managed to smile in a way that was meant to convey to anyone listening how *funny* I thought all this was. He was only gone a minute. Watching him crossing the floor to come back to the table, I couldn't help admiring the contained lean look of him. God knows what the basic attraction was. I knew he wasn't too bright, and honesty certainly wasn't a big fault with him, and yet I couldn't help liking him. In a way it was like being very fond of an adorable-looking two-year-old who keeps kicking you in the shins.

"Been talking to your Puerto Rican friend lately?" he said. I must have hesitated because he said, "Mike Ni-ve-ra," breaking it up into syllables.

"No," I said, "I haven't."

"I thanked him for sending you over to *Today* the last time I saw him." He looked at me to see what effect this remark had made.

"Thank you," I said.

He laughed in his dry, southern-raspy kind of way. "You don't think I *meant* it, do you?"

Right then and there I should have known that understanding which of Redford's constantly contradicting remarks to believe was going to be a game I'd never live long enough to learn how to win.

"What would you like to drink?" he asked me as the waiter came to our table.

"Bourbon and water," I said, desperate to please, figuring that someone from Mississippi couldn't help approving of the choice—and then wanting to kick myself when he ordered Scotch, which was what I'd really had in mind.

"Well," he said when the drinks had come, "what was it you wanted to say?"

I thought of reminding him that he was the one who had suggested the drink, but I decided to save myself the wear and tear. Instead I said, "Aren't you worried about what might happen to *Today?*"

"You mean since you and I took over?"

"Since Jim Canfield died"—I took a deep swallow of the bourbon—"since Ted Monger became editor."

"What are you fretting about that for?" he said.

I looked at him, half certain that he had to be joking. "Because he's one of Brad Rusk's boys. And because Rusk was just dying to get his hands on *Today*, except that he couldn't touch it while Jim Canfield was still alive."

"You sure do get voluptuous when you get in a sweat," Redford said. "Tell me, are all you Italian girls as sexy as they say?"

"Look," I said. "I've been working in one or another editorial department of this company for the past seven years."

"Honey, that's your big problem right there," he said, but I decided to ignore him. I said, "The thing that people like you and Wolffe and the other gold-braid characters in this company seem to overlook is *what* it is you're selling. You all seem to think it's all just advertising pages, or 'position,' or 'demographics'—whatever it is *that* means—when what you're really selling is what some editor puts into that magazine. *That's* the merchandise. And sooner or later if that merchandise gets rotten—the same as if it were apples or suits or anything else—as soon as what you're selling gets rotten, you're going to lose your customers. And, believe me, under Brad Rusk, or somebody who's being manipulated by Brad Rusk, that rot's going to set in so fast it's going to be like paper burning."

"Listen, honey," he said, turning to face me head on. "I got a little secret for you. I know you think I've been exaggerating up to now. I know you spent seven years, like you said, in edit . . . and I know how important seven years can make anything seem. But I tell you something. I been selling one thing or another just about all my life—from furniture to feminine *hy*giene deodorants—and it's the *way* you sell them that counts. And I tell you, you take the cover of any one of them seven magazines of ours, and you put it on the front of any one of them other magazines, and it wouldn't hardly make a whit's bit of difference."

I started to say something. I don't know exactly what it was because I never got a chance to say it. He had really gotten religion now. You could see it in the way he looked, the way he held himself, the way he talked, the way he didn't ever for a second let your eyes get away from looking at him.

I had never really listened to a real salesman before. Oh, I had talked to a few of them in the office, but I had always thought of them as people who had jobs and made their livings the same as everybody else except that they did it by selling.

But now, suddenly, listening to Redford going on and on, I

understood that salesmen—at least the kind of salesman that Redford was—were something else again. I understood now why, sometimes maybe six times in the same day, I'd see them go past my office, their pants-legs soaked with rain; their arms yanked down with portfolios of statistics, outsize presentation cases the size of small steamer trunks, Kodak carousels and projectors, a portable screen and various tabbed copies of the magazine; their feet moving hard and heavy, fast and frenetic, slow and deliberate, moving as though they hurt, but moving.

They sold because—whether it was suits, seduction, or magazine space—that was what they did. More important than money, or getting promoted, or even keeping the job—more important than anything was that need, the need to "make the sale."

They could quote statistics or make them up; they could use numbers to prove one thing—or the opposite; they could tell you all about the editorial content of a magazine they never read; they could slam the competition's magazine without once having opened a copy of it. What they gave the buyer was that most precious, valuable, and salable of all commodities—what he wanted. They were for the most part loud, opportunistic, pushy, and inclined to be drunk in the afternoon. But they sold. And Redford before, after, and above all things was a salesman.

"You see what I mean," he said. "You understand."

"You mean," I said, my head spinning, "that if you could do something about it—I mean, your friend Bob Moshier is a pretty big man in the company—you wouldn't do anything about Brad Rusk?"

He laughed. "You sure do get upset about nothing," he said. "How many times do I have to tell you? Rusk's not about to make trouble for anybody. He's mainly *edit*, baby—edit. He's harmless."

"But the magazines—he'll change what's going in them."

"So what?" Redford laughed quietly. "What difference does it make who decides to put in something that nobody reads anyway?" he summed up blandly. "Now . . . you gonna have another drink, or you gonna get sloppy if I let you do?" It seemed to be an academic question, even if a pretty infuriating one, because he was ordering again even before he had finished saying it.

"What's the use?" I thought, the drink beginning to take hold. I was beginning to think maybe he was right, maybe I didn't know anything. I was also beginning to think, or at least feel (which is just as bad, if not worse), that it was kind of nice being out having drinks with the publisher of one of the best magazines in the country, talking with him, discussing plans, arguing with him, even. "Now," I thought smugly (pride, as it says in the Bible, going just before you fall on your face), "*now* you're out of that front room, aren't you?"

"God, you're a funny old lady."

I thought it was one of the nicest things that anybody had ever said to me.

It was about this point that Redford decided that we'd stay for dinner. ("Never mind troublin' yourself, honey, I'll just order for both of us"—which, translated later, meant a practically raw, but great, steak.)

"You never did tell me how your Puerto Rican friend, Mr. Nivera, is doing," he said after the food had arrived.

"Fine," I said, "I guess. But you'd know about that better than I would."

"Sure I would, honey—sure I would."

I don't know whether I was more shocked by the realization of what he was saying, which hit me like the flat side of a slat, or by the fact that, simultaneously, Tom Redford was doing something once described in a book I read, published by Grove Press, as "tentatively exploring the halo of her left kneecap . . ."

Yes, I do know. Because all I had to do to get out of the second part was to move my leg and wiggle away a little on the banquette, whereas—if I read what he was saying rightly—it might take days to get over what I'd just realized.

"You mean you think—you think Mr. Nivera sent me over to you for the job because we're—friends?"

He laughed quietly. "No sin havin' *friends*," he said, all wide-open, honest, Sunday-school-eyed, but not able to keep himself from the urge to add, " 'specially in high places."

As though to water down part of what he'd said, he went on, "Look at *me*. Bob Moshier's been my buddy most of my life. Ain't even a hundred percent certain for sure I'd be the publisher of *Today* magazine if Bob wasn't head of the whole Kimberly cottonpatch." He was actually able to say it as though there were something to think about. "Old Bob," he said. "Why, we two been ass-hole buddies long as I can remember. Went to the same school . . . worked for the same bosses . . . laid the same women."

That was when I almost choked to death right in the middle of the Rainbow Room.

"What's the matter?" Redford said, slapping me on the back, "something go down the wrong hole?"

I shook my head no, still getting my breath back, and he went on, "Oh, sure—Bob's seen me with you already. Asked about you almost the first day you came. You don't have to worry none, though. He won't be bothering you any. I took care of the whole thing. I told him you were a lesbian."

"Thanks a lot."

"Listen," he said, "I don't know if you realize how much of a *saint* old Bob's been to this company. You remember a few times back a year ago when the paychecks were late—because there'd been some delay in getting here from Chicago, so they said? Listen, I tell you something. That wasn't no delay. That was just there not being any money to cover them. It was Bob that worked his tail off getting enough money to save this company." I looked at him; even for Redford conversation, it was sort of a shaking remark.

I didn't say anything; there wasn't anything to say. But just for a moment I wondered if it might not just come to the stage when there'd have to be a showdown between Moshier and Brad Rusk. After all, there couldn't workably be *two* "saints" for Kimberly, could there?

Without missing a beat—certainly without waiting for any answer from me—Redford was waving at the waiter for coffee, smiling from ear to ear, expanding, saying, "Listen, this is sure a great job. Got a whole lot of goodies going for it."

"A whole lot of what?" I said, remembering Canfield.

"Goodies," he said. "You know that trip to Paris I took two weeks ago? Sort of a fence-mending trip, seein' how some of the natives over there took exception to some of the things we wrote about them in the book. By the way, did you know them Frenchies don't like being called Frogs . . . did you know that?"

He looked at me as if he couldn't for a second understand how "them Frenchies" could have mistaken the term for anything except the most loving description.

"Anyway," he said, "I went up there to spread a little soothin' over the situation. Would you believe it, the top nigger himself met me at the airport? There was even flowers, and a band, and a police escort, and me right there in the middle—" he laughed uproariously. "Being publisher certainly beats peddlin' space."

"And that's all it means to you?"

He put me down beautifully by not bothering to put me down at all. Instead he went on, all fervor and enthusiasm. "Listen," he said, "I got some great ideas for promoting this magazine. First thing we gonna do is swing a little weight around to get it looking like it should look. None of that grub-shit paper they use for their other tacky magazines. *Today*'s special. Ought to look special. Sort of like the—the Tiffany of the Kimberly Publishing Company . . ."

"I agree about the paper," I said. "You willing to fight Chicago?"

"Don't see why not," he said. "Then there's the matter of covers. Who needs all that artistic stuff with impressionistic mountains and pictures of the sun looking like a fried egg? Sure you can have current events," he said, "people, places, things happening. Any reason why *Today* can't have them happening to a nice, wholesome, good-lookin' woman?"

"Sometimes I can't tell whether you're serious or not," I said.

"Honey, when I'm with you, I'm always serious." It was almost like a compulsive pass that even he didn't pay too much attention to. Instead he went on, all business again, all salesman. "I got about three big promotion ideas we're going to put into action right away. One of them's going to be a string of black-tie dinners. You can start working on it tomorrow. Bring in more advertising money than they ever saw."

"Black-tie dinners?"

"It's like this," he said, his enthusiasm building. "We take over some real elegant little dining room someplace—maybe in one of the fine hotels—like the St. Regis or the Plaza. The St. Regis has this room they call the Library. Real classy. Anyway, we invite us some advertising types to take dinner with us—only real high up. And maybe some presidents of companies. Or chairmen of the Board. *And* their wives. You see what I mean?" But he didn't bother to wait for an answer.

"Look," he said, "it's a real small gathering. Maybe no more than about six or eight women. And that's what we got to work on. We sugar up them fine ladies like they was prime candied yams. Bob Moshier—and Mark Post—he appeals to them intellectual type woman. Maybe even Mr. Monger—if we can get him to clean under his fingernails. And me of course.

"What we do is to give them folks a real fine social evening. No selling, you understand. Just quiet, gentlefolk conversation. And real classy food. And wine. Lots of different wines. Anyway, what we do is we send them folks home knowing they just *got* to be part of the elegant situation they experienced that evening. All those gentlemen—those top niggers—they been real impressed. But what's even more important"—his eyes gleamed in a leer "—what's even more important is—so have those gentlemen's ladies.

"And that's what we count on. Them ladies. Talking. On the way home. When they get home. When they're getting ready for bed." He grinned. He chuckled. "Why honey, don't you see what we gonna get ourselves—and just for the price of a fine dinner? What we gonna get is a genuine indoors undercover salesman in every one of them homes. Why, we can't miss, honey," he said. "It's the greatest sellin' idea since skywritin'."

The one idea—as excited as he was about it—only seemed to have started him off. Because from there he went on to tell me about the plan he had for getting some big-name writer to go on a tour of the universities and colleges—a writer whose name was closely associated with *Today*—and how that would be a great way of promoting the magazine.

"You call up that brainy executive editor tomorrow—you know, what's his name? Kreedel, the one that handles all the writers. I mentioned it to him one day last week. You go talk to him tomorrow."

And after that he had another idea. And another. And the crazy part of it was that they weren't bad at all. You could almost see the sense of his "buddy" Bob Moshier giving him this job. How it wasn't just a favor; how there could be real good business sense under all that Southern noise; how maybe, as crazy as it seemed, he might even work out to be a very good publisher for *Today*—or else, maybe, how drunk I was and didn't know it.

"Well—I guess I better be getting home," I said.

"Relax, honey," Redford said. "Don't worry—I'll make sure you get home all right. We got us time for one after-dinner brandy. What's the matter, honey? You look like you got a cramp or something. You got your period?"

"No," I said, "no, it's nothing."

Because how the hell could I tell him it was because he'd said "we"? "*We* got us time for one after-dinner brandy." Because it didn't have anything to do with him really—Tom Redford, that is. Not *really*. The fact of the matter was that ever since Steve I couldn't stand having somebody say "we" to me. A man, that is. Not if it didn't mean anything. I mean, really something. Actually, to tell the truth, it was even before Steve. It was for as long as I could remember.

"Look," Redford was saying, "why do you take all of this so serious? I don't—and God knows I got a right to. The way I figure it I'm good for this job for about two years. Shit, honey," he said, "anybody worth his salt can do anything for two years, they give him the chance to do it."

"You mean being publisher?"

"I mean *anything*. You take three-quarters of the people in high-payin' jobs—they just *waitin'* to be found out. Didn't you know that? Why, baby, you put your mind to it, you can be anything, absolutely anything, for two years, and nobody'll know the first thing about your not knowing.

"And it so happens that's exactly what I'm counting on—two years. After that I couldn't care less who's publisher of *Today*. Could be you if you wanted it. But in between that day and this I'm putting every little penny I can wring out of this job into the old tomato can. Even a few I had to stretch out a ways for. And, honey, believe me, in this particular job that's quite a few." He put his arm around my shoulder. "Like, as much as I enjoy your fine company in this fancy-ass saloon, tonight you are a genuine de-ductible tax expenditure."

"Thanks a lot," I said.

"No need to get sassy," he said. "I could have done it without ever tellin' you."

And the funny part of it was that in a million years you couldn't get mad at him. Not even when he said, "Wonder what our little Jewish friend would say if he saw us frequenting this high-costing saloon? And him trying to save Kimberly all that money."

"Who?" I said. "You mean Dave Wolffe?"

"Sure," he said. "Haven't you noticed how he's been tryin' to cut corners in the expenses? Doesn't ever hire more than *one* chauffeured limousine to drive him home to Forest Hills at any one time anymore."

Then, his mind moving elsewhere, he said, "What did you think of old Jefferson? Ain't he something else, though?"

"Who?"

"Arthur Jefferson," he said, articulating each syllable as though for an idiot child. "You did notice he was at that meeting the other morning didn't you? Or aren't you interested in chairmen of the Board?

"Boy," he said, "that's somethin'. He's about seventy-two, skinny as a lamppost, and a stiff-backed R.C. No drinking, no smoking. No screwing—at least so he says. He's even against taking booze advertising into the magazines. Took ten years for somebody to convince him to do that." He chuckled. " 'Course I ain't one hundred percent sure about that screwing part. I'd have to be real *friendly* with Miss Lorraine Clapper to really know about that."

"How long has she been with him?"

"Only about a year or so. Didn't take her too long to make an impression, I guess. Mr. Arthur Jefferson's personal secretary. Not bad for a girl that used to *sing* a little." He laughed again. "God knows, if I was paying some lady that kind of money, I'd expect a few extra amenities thrown in."

"And you haven't tried being 'real friendly' with her as yet?" I asked, the liquor obviously making it possible for me to say just about anything that came into my head.

"Baby, Jesus Christ, no," he said. "Man, I steer clear of that one. It'd be like getting into bed with a sweet, sugar-coated cobra." He took a sip of his brandy and he said, "Well, I guess I better get you home, old lady, before you turn into a pumpkin or whatever it is."

As we walked out of the restaurant I couldn't help noticing how straight he held himself. It was as though he had a dignity about him, a presence that could only have come from being dirt poor to begin with. As though having been poor gave him a way now of enjoying the "goodies" with more fun and flair than people whose bellies hadn't ever pinched them. Anyway, that was the way it seemed to me right then.

When the cab came I turned to say good-night to him, but he was getting in behind me, and when we got to the place where I lived, he insisted on seeing me to my door. That would have been fine, except that by that time he was deep into telling me exactly how to call the executive editor he wanted me to talk to tomorrow; and since it was business and he was my boss, I couldn't very well just push him down the stairs.

As we went into the apartment he was saying, "About the other thing—I think what you ought to do is to call up Miss Ginny Eisman's secretary before you go to pick up her little present tomorrow, and find out exactly where she suggests we send Miss Eisman's present to."

He had followed me into the room, and now he was quiet. I realized that this was my mother's Michigan Rummy night at Sylvia Herscher's. Needing to break the quietness, I said the first thing I could think of.

"Seriously, are you sure it's all right to spend all that money? I mean, with Brad Rusk going on his economy kick and all that."

"Rusk's no problem," he said. And then he said, mostly to himself, "It's that tall dark one that worries me. No tellin' how that one's going to jump."

"Dave Wolffe?"

He didn't say anything. I had visions of both of us standing there in that living room until my mother came home from her Michigan Rummy game, or at least until one of us dropped to the floor out of sheer quiet exhaustion.

"Would you like some coffee?" I asked.

"What I'd like," he said—the almost-lost look coming over his face, and with it, for the first time, my realizing that he was actually and deeply afraid of Wolffe—"what I'd like," he said, and then in a second, in a flash, the look of fear was gone, and it was all southern-charm grinning and the soft, gritty voice saying, "What I'd like—what I'd really like—is to have you naked in bed. How does that idea strike you?"

I didn't say. I didn't say because it didn't strike me any way; it didn't strike me any way except to leave me speechless and stock-still, quiet not out of facility or smoothness or "handling it," but quiet out of bewildered, speechless, half-smiling, scared-spitless scaredness.

In a second the moment was over—the being quiet—the smiling-saying-nothing had been enough, and all that happened was someone laughing, someone patting my behind, someone saying broadly Southern, "Well—guess maybe you got too much class for a little *poontang*," and the quiet laugh, and the door opening and closing, and the exiting of one publisher, with, *mirabile dictu*, his dignity intact for the tomorrow to come.

After a moment, I heard his steps going downward, and stupidly, with no purpose in mind, I went to the door, only at the last minute remembering to leave the chain off for my mother—smiling, terrified, thinking, "Holy dear Jesus—I guess that's what they mean by southern charm . . ."

five

THE NEXT MORNING I WENT FROM A kind of outraged shock at Redford's gall; to a kind of sick-to-my-stomach feeling that I'd probably lost my job insulting him, or whatever I had done; to a wild kind of up-and-down dipping dizziness from riding the whole nonstop merry-go-round I seemed to have won myself a lifetime pass onto.

On the way to work I went by a shop on Madison Avenue called Modes à la Parisienne. In the middle of the window they had a chartreuse silk negligée trimmed with maribou, and a pair of purple velvet slippers that turned up at the toes and had bells on the tips. Obviously they weren't things anybody ever bought for herself, and for a moment I wondered what it would be like to be somebody's mistress—somebody like Redford's, that is.

Immediately I felt as though I had been belted in the stomach. I hurried away from the window and tried to short-circuit thinking about Steve. But in spite of that, I found myself remembering one snowy Christmas dusk when we'd walked in Central Park and then run breathless up the steps of the Plaza to get warm and sat on that round little red couch with all the old ladies who always look as though they're either alcoholics or royalty or both, and Steve and I had each told the most shocking story we could think of at the tops of our voices. Leaving chaos behind us, we'd rushed laughing out into the snow and run all the way home to Steve's apartment and made love on the floor with the Bach Segovia I'd given him for Christmas on the record player and the tiny white lights blazing on the tree. Afterward, around midnight, we'd raced around the streets until we found an all-night cafeteria, each of us wolfing down hamburgers and then bowls of chili.

Enough about Steve. I made the rest of the way to work at a half-trot. As for Redford, God knows, after two months of "Don't call Steve" I might be lonely, but I wasn't *that* lonely . . . I didn't think.

Jack Sheehan came in early, put his beat-up fedora on the back of the shelf next to the filed P.I.B. reports and half-set of *Encyclopaedia Britannica* and said, "Good morning." I smiled and said hello, and—just for a change—left it at that.

In a half-daze I called Orin Kreedel's office to see if I could speak to him about the plan Redford had outlined the night before. Kreedel was in a meeting, but his secretary told me she expected him soon, and could I come at nine-thirty.

Two minutes after Jack had settled down to a pattern of staring down alternately at his drawing-board and out the window, I saw Redford come down the hallway and go toward his office. I shoved some paper into my typewriter and tried to concentrate on the script for the photography presentation I was writing. The idea for trying to get more camera advertising into the magazine was a pet of Mark Post's, and I intended to show the script to him as soon as I had it finished.

I was going into my thesis in the presentation that, since *Today*'s audience contained such a high percentage of readers with children, it was logical that it also had a large number of readers interested in taking pictures. The typewriter jiggled around as I slammed down the keys. "Furthermore, according to the most recent statistical report of the photographic industry in the United States, 59% of all pictures taken by this prospect are taken at events. That means pictures at birthdays . . . at graduations . . . at anniversaries. Furthermore——"

"That there a blackmail note you hammering out with so much vinegar?" I almost jumped out of my seat. Having finally gotten involved in what I was writing, I'd forgotten all about Redford. I should have remembered that part of his talent was always putting in an appearance the second you'd stopped worrying about his doing just that. He stepped into the room and sat down in the big brown leather chair that was near Jack Sheehan's desk.

"Well," he said to Jack, "how are things going here in the *Promotion Department?*"

Jack looked up and laughed and said, "I don't know, Tom. O.K., I guess. How would *you* say they were going?" He said it with a presence and simplicity I had forgotten about his having.

Jack's even manner seemed to rile Redford. His next comment came out butter-smooth, and deadly.

"Having any trouble staying out of them saloons?"

Jack looked straight at Redford. "No—not too much trouble, Tom."

"That wasn't a bad promotion logo you did on the November issue," Redford said. "Keep it up and you be getting *my* job one of these days." Jack laughed. He said quietly, "No, I guess not. They probably wouldn't want two rednecks in a row."

Tom laughed at that; as a matter of fact, I laughed too. But something else had occurred to me. The reason Redford was making this small talk with Jack, the reason he hadn't once spoken directly to me, the reason he'd even come into the room in the first place was that Tom Redford was even more ill at ease about last night than I was.

I said, "Is it all right if I do that assignment for you a little later this morning, Tom? I made an appointment to speak with Mr. Kreedel at nine-thirty."

"That's fine," he said, "fine. You come and speak to me whenever you're ready to go on your other little errand," he grinned. "Anytime be dandy."

I breathed deeply as he left. Evidently his ego hadn't been hurt too much. But I wondered how many times I'd have to play that rather slippery-footed kind of game called hold-off-the-boss-but-don't-hurt-his-feelings. I had a slightly sinking feeling it wasn't a game I'd ever take top money at.

Crossing over to the editorial part of the building, I felt a kind of pang, but whether it was nostalgia, or loss, or just plain hurt pride, I couldn't tell. I think the word "dispossessed" probably sums it up best. I felt as if this were really where I belonged but that I'd had the opportunity and flubbed it and now there was no chance of ever going back.

Also, another thing had happened. Because I didn't belong in this part of the building anymore, because I was no longer "editorial," I seemed to have developed a kind of over-humble, Uriah Heep attitude of which I was strongly aware and which secretly made me sick to my stomach. It was true that part of it had been picked up from the attitude of the salesmen who sold space in the magazine. To keep them from what was called "selling against editorial" it seemed that as little as possible in the way of information was given to them beforehand as to what was to appear in the magazine. It hadn't made any sense to me at first, but little by little I had come to understand what had first seemed to me almost a preciousness on the part of the editorial staff. The fact of the matter was that if the sales staff found out that there was going to be a story in the magazine that even vaguely mentioned, say, liquor, they'd use every trick they could think of to find out whether by any chance it included the name (or, hallelujah, the picture!) of Chivas Regal or Cutty Sark or Teacher's or Dewars or you-name-it—just providing that was the Scotch account they carried. If by any chance they found out that a certain brand was being mentioned, even in a fuzzy corner of an impressionistic illustration two by four inches square way in the back of the book, they'd be on the phone in nothing flat telling the client or the advertising agency that handled the client (or preferably both) that this great editorial feature "with a big picture of *your* Scotch—the label and *everything*" was being shown in the so-and-so issue of *Today* and boy, imagine if you got an ad for your Scotch in that same issue—wow, talk about editorial compatibility, talk about the right climate for your product, talk about editorial support—wow, whammy, and aren't you glad I got to you just before four-color closing date tomorrow, and by the way maybe—just maybe, I'm not sure, but

I can check, just possibly if you're lucky—if I check right this minute—there might even be a chance for a back cover—providing you went four-color, of course, and I had the order by special messenger this afternoon, that is . . .

But, whereas the salesmen took this matter of everything editorial being "out of bounds" as being completely natural (if occasionally frustrating), it was still hard for me to accept it with anything except, as I said, this pukey attitude of "here I am—forgive me for breathing."

Anyway, one of the people who most strongly brought out my feeling of being one of the crass "untouchables" was the editor I was supposed to see that morning, Orin Kreedel. To me, the few times I had seen him, or heard him speak, he summed up, as had Jim Canfield, everything that an editor should be. Added to that was the fact that, as far as I was concerned, he had one of the most eloquently forbidding faces I had ever seen. The sum total was that for the first five minutes I doubt that I gave him any real idea at all of what I had come to discuss.

I could hear my voice, sounding more and more hollow in my own ears as the seconds went by, trying to explain that what I had come to discuss was Tom Redford's plan for having a *Today* writer make a lecture tour of universities around the country—that of course *Today* would pay for all his expenses and for whatever his fees would be, that of course nobody would expect him to mention *Today* in any way—that naturally he'd be able to pick out whatever universities he wanted to speak at, and set his own schedule and take however long he wanted and talk about whatever he decided and . . . I'm not quite sure, but I think that somewhere along the line I would have reached the point of saying that it really didn't matter whether he was able to read or write, or even whether he chose to deliver his speech in English or Sanskrit or Gregorian chant, if a calm, quiet, matter-of-fact voice hadn't said, "How does Tom Redford figure he's going to make any money that way?"

Even if I could have opened my mouth, I'm not sure I could have *said* anything.

"I think Tom Redford's a fool," Kreedel said, "but the idea isn't a bad one. As a matter of fact, I have a writer in mind. Obviously I'd have to discuss the whole thing with him and his agent. He hates to come to the United States, but I think if we gave him enough money he could be convinced. And since *Today* practically discovered him and he's 'big' with the universities right now, the whole thing hangs together pretty well . . ." He stared directly at me or a long moment without saying anything, and I thought, "I'd hate ever to have to try to fake anything with him." He had a beautifully bony face and the wild kind of light-blue eyes that always strike me as belonging to a very sexy saint.

"It's all right," he said, "I don't really *eat* promotion writers, no matter what you've heard." From that moment on, at least I was able to stop acting like an "untouchable" when I spoke to him. He asked me a few more questions about the project and about the work that I did in promotion and before that, and when I told him about having been fired from *Woman's World* by Fritz Mott he threw back his head and laughed and said, "I knew there was something about you I liked. I'd hate anybody that *didn't* get fired by that fink."

"How do you like working for that fool Redford?" Orin Kreedel asked quietly as he stood up.

The question stopped me for a moment, and then I said, "I don't know. I mean—I guess it's all right."

"In a way," he said, "I guess he isn't as stupid as his super-salesman friend, Moshier."

I must have looked puzzled because he went on, "Big Bob," he said. "Now there's a *real* fool. Spends all his time out making money for the company so he can be the big hero, while all the while, back home, Brad Rusk is maneuvering to get him in the neck."

Then he smiled and sort of shrugged. "Anyway, who cares? Let them fight it out between them as to who plays God." And then more quietly he said, "Except sometimes an awful lot of good people—and things—can get mauled in the process."

Suddenly he said, "Are you keeping up with your own writing?" and for a second it didn't occur to me to doubt at all that he could read inside my head, until I remembered that, just before, I had told him about that part of me, too.

"Not right now," I said, trying in some way to make a joke out of the three words, knowing even before I said them that it wasn't going to work, pinned down almost immediately by his quiet, "When then?" which, of course, didn't wait for, or expect, or want an answer, except maybe the one you'd give yourself.

"Thank you for your time," I said, feeling again for a minute like some damn Fuller Brush salesman, except that his simple, "You're welcome," as he walked me to the door made my stupid anger just stupid, instead of something that I could have built up in my mind into a first-class grievance. ("*Nutty*" Paula, right?)

As Orin Kreedel shook my hand at the door to his office, I tried to think of how many other people in the whole Kimberly Publishing Company I could imagine ever doing that. It wasn't a large number. I wondered what he thought about Monger's being made editor, a job which, obviously, Canfield had expected him to succeed to. I wondered if I'd ever know . . .

I was about to start down the long corridor that separates the editorial side of the building from the part where I worked when a little red-haired girl with frosted bangs said, "Oh, Miss Jericho, Mr. Monger would like to see you if you have a minute."

I was surprised that she knew my name, even though we probably passed each other in the elevator or someplace else every day. I tried to remember whose secretary she had been before Monger had evidently "chosen" her—one of the salesmen's, I thought. Anyway, by now she had already gotten that real "editorial secretary" look a lot of them get for some reason or another . . . very thin and pale and kooky with sunglasses that they wear on top of their head and long straight hair and a very cool but intense look, as though they all expected to be pronounced either saints or assistant editors any minute . . .

She gestured airily toward Monger's office with a motion like something choreographed by Balanchine and then promptly disappeared. At least four seconds must have passed between the time I reached the doorway and the moment in which Monger, surrounded by what I am sure he considered all the suitable editorial accoutrements (including a blue grease pencil stuck behind his ear and, so help me, pages of manuscript scattered on the floor), looked up with phony surprise to say, "Oh, hello—come on in, muffin—come on in—sit down."

I swallowed the spit that had accumulated with the "muffin" and went in.

Actually I was grateful for the four-second delay. It gave me a chance to digest the fact that they *had*, in fact, gotten rid of "all the old man's junk." In one dizzy sweep my eyes catalogued the absence of the pictures, the trophies, the tremendous old desk. Monger himself was ensconced at a kind of pseudo-Paul McCobb, his big feet crossed on the table in front of him; his tight-muscled shoulders slung so deep in a reclining chair that his back was almost parallel to the floor.

"You wanted to see me?"

"Yes," he said, "I figured since you were going to be working for me now I'd better get to know everybody."

"I thought maybe you'd started firing people already," I said, airily, if somewhat suicidally. I knew that that was the last thing that Monger wanted to hear said out loud—the image of being some kind of a hatchet-man on Brad Rusk's behalf, the implication that he wouldn't or couldn't work with Canfield's tight little staff, all of whom knew more about magazine publishing than he knew about keeping his face clean. I knew that that's what the comment would mean to Monger, and how it would endear me to him from here on in. That was probably why I said it.

He sat up so fast I thought for a moment his chin was going to hit the table. It didn't, and he said, "Christ, don't say *that*." At least it was an honest reaction. What wasn't even two cents' honest was the smile—at least I guessed it was supposed to be a smile. What it looked more like to me was the slitty mouth on a stuffed shark I had seen once in the Museum of Natural History, a place I used to spend a lot of time in when I was a kid. I got so hung up on thinking about those old times that Monger was at least two minutes into his little speech before I caught onto what he was really saying. He was talking about Tom Redford's small, executive dinner idea which I hadn't realized he already knew about. "I don't know what kind of little box-supper 'do's' they may consider pretty fancy back in Mississippi where Redford comes from. It wouldn't surprise me if he expected to have the women wear gingham. Tell him I said no."

Before I had a chance to answer, he said—deadly quiet—"What were you seeing Orin Kreedel about?"

Shaken by the inescapable attack of the question, I *willed* myself to show absolutely no reaction. Instead, as briefly as I could,

I told him about Tom Redford's plan to have a famous writer that *Today* had helped develop make a tour of the universities. "It's a banal idea," Monger pronounced, even before I had finished the last sentence. "Anyway, I know the writer Kreedel has in mind. He's a cliché. Wouldn't get enough kids to fill half the auditorium. If Kreedel kept his thinking a little further up to date than last year's copy of the *Sewanee Review* he'd have known that. Well . . ."

He stood up, and I was so surprised by the movement that it was two whole seconds before I realized this extravagant gesture on his part was meant to convey, That's all, kid, it's been nice to see you—now get lost—*beat it.* He was still smiling as I left his office—or whatever the slimy expression was that passed for a smile as far as he was concerned. I don't know what made me angrier, the way he smiled or the sudden realization that in fact what he had done was to laugh at, negate, and step on every single project that Redford had given me. The more it sank in the madder I got, so that I was still shaking with anger when I got back to my office. It was almost ten-thirty, and the first thing that Jack Sheehan said as I came into the room was, "Tom Redford's been looking for you." "What for?" I snapped—my usual sweet self.

"He said something about a 'liddle ol' errand' you were going to go on for him."

Jack's occasional imitations of Redford being "southern" usually come out so thick you can't tell what he's saying, mostly because having come from Arkansas himself, he can never stand what he refers to as "Redford's professional spoon-bread talkin'."

"Oh, damn," I said. I got up to go into Redford's office, but I was still shaking so much that Jack noticed it.

"What's the matter?" he said. "Something happen?"

"No—no," I said. "Nothing." Nevertheless, I sat right down again; late or not, I wasn't about to go right into Redford's office in the state I was in.

What the hell was I getting so excited about, anyway—it wasn't as though it really had anything to do with *me.* Just because I happened to think Redford had a nice way of wearing his clothes, and just because his bones seemed to hang from his shoulders with a certain flair, what did that have to do with me getting furious with Monger if he wanted to make a fool out of Redford? The fight was between Monger and Redford. God knows they didn't need my help. All I was was a little promotion writer—and promotion writers got paid to write promotion—and that's all. Especially this one, I thought.

I took one more deep breath and started toward Redford's office. Thank God I'd caught myself in time, I thought—God knows what nutty thing I might have done otherwise—got myself in the middle of some fight that didn't even have anything to do with me. Thank God I'd caught myself in time—that's all I could think, just thank God. At which point in my soliloquy I reached Redford's door, stomped into the office, slammed my hand down on his desk and said, "Who the hell does he think he is anyway? I don't even *work* for him!"

I'm not at all sure at what point Redford began to make any sense out of what it was I was saying. All I know is that I spilled out all of my fury about Monger in front of him the way you'd empty a bag of peanuts, and that even after it was all emptied I still went on saying, "If he had something to discuss with you why didn't he discuss it with *you?* Doesn't he know you're the publisher? Who does he think he is, anyway? An *editor!* He's no more an editor than—he wouldn't even *recognize* an editor—not if he tripped over one——"

"You finished now?" Redford said. Quietly. So quietly that it stopped me dead.

"Yes."

After a minute he stood up. Then, finally, still very quietly, he said, "I'm going to say something. And I'm only going to say it once. We both know who's running this magazine, don't we?"

"Yes," I said, my voice almost as quiet as his had been.

"Then go ahead. Do your work," he said, suddenly as fast as a cat around the desk and patting my back, all smiling and charm and "pick out somethin' pretty now—don't forget!"

He put his arm around my shoulder and I said "damn" to myself, feeling my stomach flip and my toes curl at the same time I saw him to be the professional charmer and con man that he was. Nevertheless, I couldn't help feeling my blood running clearer as I walked down the hallway before starting out on my little *objet d'art* expedition. Even the fact of how completely I'd betrayed my own decision of no more than five minutes ago couldn't dampen me completely.

Then suddenly, as I was putting on my coat, something in Tom's way of handling my outburst that morning struck me as being—I didn't quite know what the right word was. "We both know who's running this magazine, don't we?" he had said. And I had said yes.

But the fact of the matter was I didn't know *what* I knew. I didn't even know what it was he *expected* me to know. The complete fact was that I didn't really know what we were *talking* about.

In the meantime, thinking this, I had buttoned my coat up against the snow that had started again that morning, gone down the elevator, and started over to the little antique shop to pick out "something pretty." It wasn't until I pushed open the door half an hour later that I realized I'd never bought anything that cost twelve thousand dollars before . . .

Later that day, after the purchase had been made and I had returned to the office only to be told by his secretary, Jane Perry, that Redford had been called to a meeting with Dave Wolffe and Brad Rusk, I found my anger at my involvement in this whole mess coming back in fast, building waves. In spite of being able to say to the little blue-slippered antique dealer, "Charge it to Mr. Thomas Redford, publisher of *Today* magazine, please," as I handed him one of Redford's lovely little cards, and in spite of his not batting an eyelash at the whole transaction, I still didn't *feel* like somebody who could walk in off the street and order something for twelve thousand dollars without even asking to see it. It just didn't feel right. What it did feel was fishy. Which—since it *was*—made me all the angrier.

I kept trying to remember that one of the original reasons for Redford's hiring me had been that I had so much "class," but that didn't begin to make even a crack in which I really felt like, which was some snot-nosed little kid walking into Lord & Taylor for the sole "purchase" of one free shopping bag, a comparison I hadn't made up strictly out of my imagination.

At any rate, the more I thought about it, the angrier I got. As the afternoon wore on I found myself repeating the speech I had made to myself that morning, "Let *them* get involved if they want to, God knows they're getting paid enough for it . . . what does it have to do with me anyway . . . etc., etc., etc., at which point, I think, I almost expected Redford to walk into the room and kill my whole resolution, as usually happened when I got myself way up to this pitch. Except that this time Redford wasn't there; as a matter of fact neither was Jack Sheehan. Actually, the only one I had any dealings with at all that afternoon was Pete Larsen, and there certainly wasn't anything about *him* that contradicted my resolution about Kimberly just not being worth the trouble.

Right at the dot of five (Patty wafted past in a sad cloud of Arpège, headed God knows where . . .) I yanked on my hat and coat, and—suddenly desperate to get away from the place—I shoved my way into an already-packed elevator.

In the middle of supper my mother smiled and said, "You didn't call me today."

"Gee, I'm sorry—I got a little busy at work."

"Oh . . ." and silence.

"Why didn't you call *me*, Ma? It's easier that way."

"I never like to do that."

"Why not?"

"Because you might be busy."

Sometimes it makes me feel better, knowing that I inherited being a kook. As somebody once said about my mother, "She's so great she makes you forget she's a Catholic."

Right after I finished dinner and helped with the dishes I went right inside to the desk, sat down and changed the ribbon on the typewriter. Sharpening pencils, changing ribbons on typewriters and going through old, unsold (but naturally salable) manuscripts ("as soon as I have a chance to make a few changes in it") is standard writer's therapy. The more I bustled around that typewriter, changing the ribbon, trying out the red half, opening the top and dusting the keys, including poking out the junk that always gets stuck in the lower case "*e*'s" and "*o*'s," the more convinced I was that all the Kimberly nonsense had gone on long enough, that I had wasted enough time and that, even if I was a kook, I was still, funny as it might seem, a writer too . . . Almost as though some invisible doubter had said, "Who says?" I found myself pulling open the desk drawer and scrounging among my collection of little telephone books, fourth-class mail I have trouble throwing away, assorted paper clips and "notes for things" that are in fact notes for things, and a lot of old junk, until I came up victoriously with a clipful of old newspaper reviews on my book . . .

I plunged deep into the lovely collection. "This first novel reflects great promise by its young and attractive authoress . . ." (I sat straighter in the chair, promising myself to lose five pounds the very next week.) "Miss Jericho undertook to exploit a powerful and often difficult theme. This she accomplishes admirably . . ." "Here is a writer of truly remarkable promise . . ." Suddenly what had started out kind of tongue-in-cheek took over in absolute earnestness. "A first novel distinguished by rhythmic prose and excellent choice of detail." "If you don't know what real desperation is like, you should read this book. It will make you more compassionate . . ." "Beautifully written . . ." "An unforgettable novel . . ." "Moving . . ." "Full of warmth and compassion . . ." "Outstanding writing . . ." "Admirable . . ." "Strong . . ." "Vital . . ."

Angry, inspired, impatient—almost in fever-pitch, I pulled open the bottom desk drawer, scattered all the junk onto the floor, found and read the last struggling attempts at a short story I had started a month ago, jammed the paper into the machine and started to type. What came out was:

TO: THOMAS E REDFORD, PUBLISHER, AND MARK POST, ADVERTISING DIRECTOR.

FROM: PAULA JERICHO, PROMOTION STAFFER

(At first I thought it was a joke. I felt myself typing it, but it was a joke. Except I *did* keep typing on . . .)

"People are always talking about the 'spirit' of *Today* magazine. When I first came to work here I thought it was just another made-up promotional phrase—like a kind of corny gung-ho thing . . . Like the way companies try to make you feel like a "team" or something, instead of giving you raises, or something like that. But after a while I saw it was something else, something *real*." I kept on typing, hardly aware of anything—only the words coming faster and faster. ". . . anyway, all of a sudden, tonight, I wanted to say thank you to the two people who it seems to me keep that ghost alive. When Jim Canfield died, and when the other things happened right after, I think most of us, even if we didn't say it, felt just about the same way. Scared, I mean. But at the same time I think we felt sure that *Today* wouldn't really change all that much. And I think that was because of you two. I guess it might look presumptuous of me to say these things, especially seeing as how I'm new to this job. But I don't think that's important. I mean, how it *looks*." I kept banging away, banging and banging, until finally I found myself slowing up.

"Well, I didn't mean to take so long saying this. 'Thank you' is just two short words. And that's all this is really about. So—'thank you.' "

I think I heard the typewriter stop before I was aware of not writing anymore. I looked down at the pages. I had no idea of what I had written, but I had more than an inkling that it wasn't completely within the bounds of Redford's lessons in "knowing my place." I was exhausted. I looked down at the unfinished story lying next to the typewriter. The struggled-with sentences stared up at me like some damn cat I'd forgotten to feed. On the other hand, on the other side of the typewriter were the two tightly filled sheets. Two pages—and hardly a stop. Slowly I picked them up. Holding them, I hesitated a second, stopped, smiled. Then I patted the shiny black flank of the somehow accusing-looking typewriter. "Maybe *tomorrow*, baby—O.K.?" And I started to read over what I had written.

six

WHEN I WENT INTO THE KITCHEN THE next morning I found a note my mother had left with the coffee cup and paper napkin she puts out for me every night. It said: "Dear Paula, Please bring some rubber bands. I always say a prayer for you Honey. Love, Vivian. Excuse writing as I cannot see."

The first thing I did as soon as I got into work was to make a Xerox of the letter I had written the night before; then I put one on Tom Redford's desk, the other on Mark Post's. For a second after I had done that I had the sinking feeling that maybe it hadn't been such a great idea after all. Anyway, there were too many things for me to do that morning to spend much time worrying about it. Already the phones had begun to ring.

I was just getting back to my desk when Pete Larsen came into the office. It seemed to me that every day you could see him get smaller and meeker and more harassed by Redford; yet there was something about him you had to admire. He might very well drown under Redford's direction, but he'd do it with dignity. Rabbity dignity, but dignity.

What Pete Larsen wanted me to do was to make what are known as "fact sheets" out of other things that are known as "Starch reports." "Fact sheets" (as I had finally grasped after several weeks of absolute certainty that I would never grasp anything about them at all—much less ever be able to "make" one) were eight-and-a-half-by-eleven-inch yellow pages with headings that said things like:

TODAY READERS ARE MATURE, AFFLUENT AND INCLUDE A HIGH PERCENTAGE OF COUNTRY CLUB MEMBERS—or

TODAY WOMEN READERS ARE AWARE, WELL-EDUCATED AND OWN AN AVERAGE OF 15 PAIRS OF SHOES (NOT INCLUDING SNEAKERS)

Sometimes they made comparisons between us (meaning *Today*) and other magazines—or what is known professionally as "the competition."

TODAY HAS MORE PASSPORT HOLDERS THAN THE NEW YORKER, ESQUIRE OR HOLIDAY—or

TODAY HAS A HIGHER PERCENTAGE OF MALE READERS REGULARLY USING MEN'S TOILETRIES THAN THE SATURDAY REVIEW, PLAYBOY OR SPORTS ILLUSTRATED

Of course names of other magazines (the "competition") kept changing on the different sheets. Sometimes it was *The New Yorker* we were better than—or sometimes *Sports Illustrated*, or *Esquire*, or *Holiday*—once we even had a sheet that showed how much better we were at "delivering" readers in the twenty-eight-and-over age-range than either *Jack and Jill* or *Humpty Dumpty* magazines. The trick was, of course, to use whatever other magazines you had to that would prove your point. Proving your point is very big in promotion (magazine-advertising promotion, to be more exact). However you have to do it, proving your point is the thing to keep your eye on. Logic is nice, but not necessary.

What all these fact sheets were "made up from" were statistical findings known as "Starch, Simmons, or B.R.I." And, actually, I guess all this made *some* kind of sense. But you couldn't prove it by me. The closest I could come was by thinking of it in terms of some organization "proving" that, say, redheads were sexier than blondes—or that men under five-feet-six were better in bed than those over five-feet-six. It was as though, once this factor had been established by a "disinterested" source (like a "Starch," say) any guy under five-feet-six could begin making his pitch with this point (in this case his "performance factor") already established.

The funny part of it was practically all the magazines went along with this game—even when it didn't help their cause at all. Like, for instance, the year Starch began to "prove" that more and more women were buying liquor ("booze" in magazine parlance) and *Today* had a very low percentage of women readers. Instead of just ignoring the whole thing, the liquor salesmen for *Today* started some double-talk about *Today* men having taste that was just as fine and discriminating as *Today* women, until one day some wise guy in one of the advertising agencies said, "Are you trying to tell me *Today* has a high quota of fags?" and the salesmen dropped that particular tack.

The wildest part of the whole thing was that the magazines *paid* to be part of these surveys. As if, no matter how detrimental it was for them, they still couldn't afford to be left out. In a way it was like a magazine Mafia.

I listened to all the instructions that Pete Larsen was giving me, and then, as soon as he had finished, I said, "Fine—I'll get started on them as soon as I've finished the projects I have to do for Tom." I watched him gulp, not feeling the least bit guilty, even though all the while it was supposed to be him—as promotion director—that I was responsible to. Anyway, the way he looked at me, I think he felt it was probably something dirty.

It was almost ten o'clock; I bunched all the stuff on my desk into a neat-looking pile and got ready to go over to the editorial side of the building. A photographer was coming to take one or two pictures to include in my photography presentation. I thought of them as "editors-at-work." It was supposed to prove that because our editors used such fine photography, naturally our readers took a lot of pictures too. It was the logic-is-nice-but-not-necessary thing all over again.

I had just come from the ladies' room—bracing myself for the inevitable encounter with Monger, hoping hopelessly that maybe he'd be out of town and we'd have to take the picture without him—when I ran into Redford. I'd already grabbed up my pencils and notes and things, and for a moment I thought all I was going to have to do was say, "Excuse me," and sidle past. But that was pure fantasy.

"Come in my office," he said.

Deliberately rude, he stomped into the room ahead of me, went behind his desk, grabbed up a piece of paper that was lying there and demanded, "What's all this crap supposed to be about?" It was the letter I had written the night before.

"You haven't done anything with it, have you?" he demanded.

"What are you talking about?" I wasn't exactly screaming, but on the other hand I'm sure they didn't have any trouble hearing me forty yards down the hall.

"I mean you didn't put it on the bulletin board or send it around or anything?"

"What's the matter with you anyway?" I yelled. "Can't you read? It says it's to you and Mark Post. Why would I circulate it around, or whatever it was you said? I had something I wanted to say—to both of you—and I said it. Didn't you ever get a memo before?"

"Oh," he said quietly, sitting down. "It was just for the two of us . . ."

"What did you think?" I asked. "Did you think it was some kind of promotion piece? Or maybe you thought it was for an ad in the *Times*. Or maybe you thought . . ." Except suddenly I was petering out. I realized that what had really made him yell was something as nutty as embarrassment—awkwardness, shyness even, over what I had written. It almost seemed to be catching. Because suddenly there was something almost timid in the awkward crudeness with which I finally sputtered out, "Didn't anybody ever say anything *nice* to you? What kind of people are you used to, anyway?"

He was absolutely still for an entire minute. Then, very quietly, he said, "You're right." If there was anything that could make you feel more in the wrong than Tom Redford yelling at you, it was Tom Redford apologizing. "I guess I just don't know how to handle something like that," he said. It was the sum total of all the abjectly humble apologies ever made. I found myself fumbling with half-sentences that started out, "It's all right," but I remembered how furious I'd been a minute before (and how outrageous he had been), so that by the time I left his office a minute later, all I knew was that in the space of about three minutes, I had been more deeply infuriated and charmed (read: *conned*) than I'd ever been in my whole life. "The damn bastard," I thought to myself, at the same time smiling (and dizzy) at the stunning swiftness of still another Tom Redford instant emotional-about-turn. But before two minutes had passed I had completely forgotten Redford and the events of the morning so far.

The free-lance photographer was waiting for me when I got over to the editorial department. He was a little red-haired man who seemed completely overpowered by his equipment—to say nothing of Ted Monger, who had already pounced on him by the time I arrived and was already deeply immersed in "planning the shots." (At one point he actually formed a camera "square" with his two hands, describing the "angle of depth" that he thought would be most "visually perceptive.") The most interesting thing to watch during this whole time were the faces of Canfield's staff, the people who had planned and executed the great journalistic strategy that had been *Today* for so many years. I especially enjoyed watching Orin Kreedel, the executive editor. I had never before seen a person capable of delivering so much eloquence without saying a word. Wherever Monger moved in the room, Kreedel kept his eyes completely on him, and yet he did it without ever seeming to care enough to bother. Kreedel's glance moved over and absorbed Monger, heavy at the same time with both judgment and dismissal. It was fascinating to watch a man who could at the same time be so intense and so cool. Once, when Monger was rearranging his "group" for the fourth time, Kreedel said quietly, "Wouldn't it be better if they just used a picture of you alone, Ted? Wouldn't it be more forceful that way—you know, the sole, strong, guiding hand at the helm of *Today?*" And for a second you could see Monger actually considering it, taking it straight, and then deciding that no, it would be more authentic "with my helpers around me . . ." So help me, he said that.

By the time the photographer had finished, I was a mass of quivering something-or-other—the main ingredient of which was a kind of spitting-mean fury at Monger. There was something about his smiling-shark one-upman conceit that made me bristle just being in the same room with him.

Finished with the photographer, he turned to me in a kind of half-gesture of dismissal. "All right, Paula, make me famous." To make it even worse, Brad Rusk strode into the room just then, his wild, large-pupiled eyes making him look like a drug addict.

Monger said, "Hi, Brad—you should have come sooner. We could have taken some pictures of you."

"What's it all about?" Rusk said.

"Just some editorial shots for a presentation that promotion's putting together," Monger said. And then laughed. "One of Tom Redford's brilliant little ideas, no doubt."

Rusk looked at me and said, "Oh, that's right. You work for Redford, don't you?"

"Yes, I do," I said.

"Redford," Rusk said, using the name like a curse. "Another one of Moshier's high-paid puppets . . ."

I could feel the anger rising in me, and I forced myself to swallow it.

Immediately, with a kind of airy, dismissing wave of the hand, Rusk said, "What the hell, they're a *temporary* evil," and strode, Caesar-like, out of the room. A moment later Monger strode out after him.

A temporary evil. He had actually *said* it—in so many words. And nobody—*nobody* had asked what it meant. The thing was that nobody seemed to question it, *even in their own heads*—if you know what I mean. It was wild.

Finally, with the whole nonsense finished, I thanked the editors and the photographer and left.

I was about to start down the long hallway toward the business end of the building when I saw Eileen Stevens, a tiny, red-headed woman, hurrying after me. Eileen is about a hundred and four years old and used to be on "the English stage" doing, as I had once heard it described, "comic turns." As she came down the hallway toward me, her arms flapping, her bright blue cardigan hanging loosely from her shoulders, she gasped, "Paula—Paula, how are you?"

"Fine," I said, "just fine. How've you been, Eileen?"

"Oh, just fine, but you know—*busy*." Her eyes rolled skyward and her arms flapped upward once again. I tried to think of something else to say to her, but I couldn't think of anything. Finally I said, "Pretty hectic, huh?"

"Oh—oh, you can't imagine how it is with *him* . . ."

I tried to remember who it was that she worked for but it escaped me. Finally I decided I'd had enough of it and said, "Well, Eileen, take care of yourself. I'll be seeing you," at which point she practically went into a fit, evidently panicked by the thought that I might get away before she had said what she had run after me to say.

"Oh," she gasped, all breathless, panting, "Oh—Paula—would you—would you have a minute for Mr. Wolffe?"

Suddenly I remembered. *That* was who she worked for. Dave Wolffe. Chairman of the Magazine Division. The newest monster of the Kimberly Publishing Company. Or so everybody said. Suddenly I thought of Tom Redford . . . "it's that tall dark one I'm afraid of—no telling which way he's apt to jump." "Yes," I said, as cool as cream. "Yes, of course."

She babbled all the way back to the carpeted area past editorial where evidently Dave Wolffe had his office. I had never been in this part of the building before. It was divided into large, lush executive offices, each with a secretary seated directly outside who seemed to have no other function except that of watching me being led down the hallway to Wolffe's office. I did my best to make the performance worth their while.

Fluttering all the way, Eileen conducted me right up to the door of Wolffe's office, only to stop with breathless abruptness.

"Oh—*oh*," she said, practically barring the way. "Mr. Wolffe's —Mr. Wolffe's on a long-distance call. Would you—would you mind waiting?"

I stepped back with all the *noblesse oblige* I could manage, smiling coolly, projecting "serenity" right on.

"That's the way it is all the time—he's always so *busy*," Eileen informed me, and then, flustered, "I'll get you a chair. There should be a chair out here. I keep saying there should be a chair so people can *wait* when he's *busy*——"

"No, really," I said, "I enjoy moving around." The act I was playing, I was half surprised not to hear myself use the royal *we*. "We enjoy moving around—yes, indeedy do, we do . . ."

"Well—well, if you're *sure*," Eileen said.

I smiled graciously. I was aware of the other executive secretaries interrupting, as I said, their doing nothing to take careful note of me. I had a feeling that they knew what color underwear I was wearing. I stood splendidly. Occasionally I took a step or two—splendidly. I stood the way you do when you feel as though your dress is really a *robe*. I have a vague impression I probably sucked in my cheeks. I might even have pouted a little. God knows, I probably would have gone through all ninety-four Actor's Studio expressions if Wolffe hadn't gotten off that phone in what was probably two minutes at the most. All I knew was that I'd have willed myself to drop dead rather than show one half-inch of the bewilderment I was feeling. He was still standing next to the phone, unsmiling, noncommittal, cold-eyed, as Eileen fumbled me into the office . . .

I'll never forget the scene. It was as if the script directions read: [*With equal archness on both sides, played fast, cool, and to the point.*]

Wolffe: [Without preliminary] Do you think we could have lunch?

Me: Yes.

Wolffe: Twelve-thirty all right?

Me: Yes.

Wolffe: Monsignore?

Me: Fine.

Wolffe: You know the address?

Me: Yes. [*Quiet, cool, elegant-to-the-teeth exit. To be murmured under her breath, inaudible but indicated*] Bastard—"Do you know the address?"

Back in my office the first thing I did was look up the address. It wasn't until I had actually jotted it down that the full bewilderment of what had happened washed over me.

It was eleven o'clock now. By twelve-thirty I had to be at the restaurant, and I didn't have a clue to what the whole thing was about. There were all sorts of little fragments of feeling. I was confused, bewildered, excited, intrigued, over my head—and scared. The thing was, when you put them all together, what they came out as was *angry*. The monologue went something like: What the hell does he think he's doing, anyway—who does he think he is—what makes him think he can just come in and take over the place, et cetera, et cetera, et cetera. The thing was that it was obvious—even to me—that the more "angry" I became the more scared I really was.

Finally I barged down the hall to Redford's office, but he wasn't there. I went into the little room that Jane Perry used, and I said, "Can you tell me where Tom Redford is?"

She looked up at me, friendly but all business-like, the way she was, and said, "Oh—well, Mr. Redford had to go home. He had a sore throat."

No wonder the South had lost the war, I fumed to myself. "Well," I said, "all I know is that Mr. Wolffe has asked me to have lunch with him. I wanted to ask Mr. Redford what he thought about it."

"Oh," she said, "in that case I think you should definitely call Mr. Redford at home." I couldn't help admiring the cool, efficient but uninvolved way she handled things. "Here's his home number," she said, writing it down in small, precise numbers on a piece of white paper. I thanked her and went back to my office.

As soon as he picked up the phone and I heard his southern, "Hel—lo," I plunged right in. "Tom, this is Paula Jericho. I'm sorry to disturb you at home, but I just wanted you to know that Dave Wolffe just called me into his office and invited me to have lunch with him today, and I just wanted to talk to you to find out what it was all about." I said the whole thing in one breath and probably I would have repeated it all over again, except that I was suddenly aware of Redford laughing.

"What are you getting so excited about?" he said.

"What do you mean?"

"He just wanted to have lunch with you because you're a sexy girl, is all." He was still laughing.

"All right," I said angrily. "But don't say I didn't tell you," and hung up. It was just twelve o'clock. Counting fifteen minutes for the ladies' room and ten minutes to get to the restaurant, there were five minutes left before I had to do anything. Suddenly I was positive that Wolffe really wanted to ask me about *Woman's World*, the magazine I had been fired from. He probably had had time to find out what a fool Fritz Mott was. Maybe he even wanted me to go back to *Woman's World*. But who did he think he was anyway? What made him think he could just come in and *buy* the whole operation? . . . And on and on, slipping back into my original litany—with one variation. This time I promised myself that no matter what he said, I wasn't going to tell him what was wrong with the company, or how to save it, or my ideas, or—anything.

This time I was going to keep my mouth shut even if it killed me.

Pace.

seven

EVEN WITH MY EXPENSE-ACCOUNT TRAINing on *Woman's World*, the restaurant was one of the plushiest I'd ever seen. The entrance was one step up from the street through a revolving door which an eight-foot-tall doorman made a great fuss about helping me to turn. "O.K. then," I thought, seeing which way it was that things went here, abandoning myself completely to an absolutely helpless little-me performance that ultimately ended with four different waiters assisting me with my coat, picking up my gloves, tucking me into a banquette behind a white tablecloth and purring into my ear about how Mademoiselle would have a drink now or wait until monsieur . . . ? I allowed as how I would "wait until monsieur." Wolffe was fifteen minutes late, and for some nutty reason I got the idea he had planned it that way. I watched him stop to chat with the headwaiter as he checked his coat and—summoning every ounce of willpower I had—managed to be the picture of aloof but pleasant composure as he reached the table.

"Hello. I hope you haven't been waiting long."

"Oh, no," I said. "I just got here." *I wouldn't give you the satisfaction.*

"Have you ordered a drink?" he asked me.

"No, I haven't."

"What would you like?"

"Scotch and water, please." I wasn't about to be caught in that trap a second time.

"Campari and soda for me," he said to the waiter, and then, to me, "Have you ever tasted it?"

"No."

"I learned about it while I was in Italy," he said. "It has a wonderful taste. Sort of dry and—interesting. I think you'd like it. Why don't you try it?" he said, detaining the waiter with a look.

"No, thanks," I said, smiling, "I'll just have the Scotch and water."

"Make it Dewar's," he said to the waiter, and I thought: Well, Paula, you've finally met him—Mr. One-upmanship himself.

"I still think you'd have liked it. If you have it with gin and vermouth, the Campari, that is, it's called a negroni. It was Hemingway's favorite drink."

"Yes, I know," I said. *My God, now he's going to be* literary.

"Have you ever been to Italy?" he asked.

"No."

"I loved it," he said, "especially Positano. I'd drive down this long, winding road in a convertible going ninety miles an hour cursing in Italian all the way, and with my coloring and the way I sounded, they took me for a native, and they'd curse right back at me with the wildest profanity you've ever heard. It was wonderful."

His skin *was* on the darkish side, and actually his eyes lost some of the coldness when he smiled, but I couldn't help thinking how much I hated men who had that kind of *dry*-looking brown hair.

"I'm Sephardic on my mother's side."

I'll never be sure that I didn't literally jump back, especially since I didn't quite know what he was talking about.

"It means that she was a descendant of the former Jews of Spain. It's a special kind of thing. That's why I'm so dark," he said.

He looked almost shy for a moment, and then I remember thinking: *shy as a shark.*

"You should go to Italy," he said. "You would like it there."

"I'm sure I would," I said, "but I don't know about coming back."

"What do you mean?"

"Oh, I don't know. It's just a feeling I always had. That if I went to Europe—to Italy—I wouldn't ever come back. I don't think I'm really *American* anyway," I said, surprised to find myself saying all this.

"No, you're not," he said, and somehow I resented him knowing this thing about me so quickly and with no need for explaining.

"Were you in Europe recently?" I asked.

"Just a few months ago. Right after I came to Kimberly. I called it business, but it was really a boondoggle." He smiled, admitting the whole thing—daring you to dare to have an opinion.

"Did your family go with you?"

"Yes—all of us. My wife and the two kids. I have two boys." He grinned, "They're kind of nutty kids. But bright." *Naturally.*

Just at that moment, Rudolph Hecht, *Today*'s roué-type public relations manager, entered grandly, escorting a young, brown-haired girl who must have been, at most, one third his age. I recognized her as one of the new secretaries on *Woman's World;* I doubt whether she'd been there more than a week.

It was strange—and funny—to watch Hecht's grand entrance just sort of collapse as he noticed Dave Wolffe sitting there. There was an evident impulse on his part to turn around and leave, and even though he kept himself from doing that, I could see how blanched his skin looked as he nodded to us when they passed our table. Then there was also the hurried little discussion and the small mutterings that went back and forth until finally the headwaiter understood that what Hecht was asking for was a table as far away from us as possible.

After it was all over, Dave Wolffe laughed quietly. "Wait until he tries to get that passed on his expense account." And I realized how right Tom Redford was not to underestimate this one.

"You're not married, are you," he said.

"No, I'm not."

"Were you ever?"

"No—no I wasn't."

"You have a boyfriend, though."

"No—I mean, not anymore—I mean—" I finally stuttered my way into silence, furious at having told him that much.

"How come, I wonder." It was more like his feeding a question into his own brain—like a computer almost—than it was that he expected me to give him an answer. Anyway, this time I managed not to volunteer one. He went on pondering his self-put question for a few more moments, and then, out of the clear blue sky, he said, "Are you afraid of me?"

He had about the most piercing look I had ever seen; even the nuns I had known couldn't top him.

"No," I said, surprised.

He didn't pursue that one any further. At least not right then. Instead he nodded sort of half-grudging, admiringly, and said, "I've got to give you credit. You never batted an eyelash when I invited you here."

That'll be the day, I thought.

"Weren't you surprised?"

"No. Should I have been?"

"Have you ever been here before?" he asked, and when I admitted I hadn't (one of my few concessions to honesty for the day), he said, "I have some money in it. Sometimes I come here three times a week." It was wild. For a second I almost went crazy trying to put my finger on what it was he was reminding me of, and then all of a sudden, right there, I got the picture—Jimmy Powers. Jimmy Powers had been my beau in the fourth grade. Once he had jumped the empty space between the school roof and the next building. When he had gotten safely to the other side he had turned around and yelled to me where I stood sick to my stomach afraid, "Hey—did you see me? Did you see?" Just for a split second it had been like Jimmy Powers all over again.

Anyway, in the very next second, the illusion had gone. He

was saying, "Brad and I are going to Washington tomorrow. Kimberly's giving a big dinner for some advertisers. Tony Randall's going to be the M.C. Do you like him?"

"Yes, I do." *As though, if I said no, the whole thing would be changed . . .*

"I saw that new movie of his last week," Wolffe said. "I think he does Madison Avenue comedy better than anybody."

I nodded to agree with him, except that all of a sudden, almost without any breath between it and what he had been telling me, he said, "You're sure you're not afraid of me?"

I said, "No. Do you want me to be?"

"God, no." He grinned broadly and dropped the subject, permanently this time, I hoped.

"Have you always lived in New York?"

"Yes," I said, "as a matter of fact, at one time or another we've lived in every borough" (wondering immediately why I had said *that*).

"It must be lonely," he said.

"I beg your pardon?"

"Your not being married."

I remember reaching out and taking hold of the glass and swallowing about the longest drink of Scotch and water anyone's ever managed, the point being that as long as I had the glass to my face I didn't have to answer anything. Anyway, after at least a half-minute I knew I couldn't keep holding that damned glass to my mouth much longer without choking to death, so I said, "Why do you say that?"

It seemed like a workable enough party to me, except that almost as soon as I had used it he said, very simply, "Well, isn't it?" and I was right back where I had started from.

"When I was four years old—that summer, when I was four, I was sure I was going to die of being lonely. I even buried my marbles under a tree in the back yard. I didn't want my brother to get them after I died. Some sweet kid, huh?" He grinned again, like some god-damn lonely four-year-old, and I said to myself: You're a rotten, no-good, know-nothing from some stupid construction company, who knows absolutely nothing about magazines, and you're so stupid you don't even know that Brad Rusk is a big, fat, stupid slob . . . and then I could hate him again, nice and easy the way I had planned to from the very beginning.

Why would anybody in the world let himself be so *thin*, I wondered, concentrating on that while he sort of kid-school grandly got the waiter's attention and asked for the menus. Crazy, dry-looking, sort of brown hair. It was crazy. It wasn't arty. It wasn't mod. It wasn't beat. It certainly wasn't good-looking. The most you could say, if you wanted to say anything at all, was that it was "neat."

"I keep in shape by fencing every weekend. I can't stand fat men."

The waiter handed each of us an immense menu, and Wolffe said to me, "My sister's the good-looking one in the family." And I thought, Paula, old girl, don't try to grasp any of this right now. Just keep your cool and smile and look at the stupid menu, and what if it does come out that every damn thought in your head is written in eight-foot letters across your forehead . . .

"You used to work on *Woman's World*, didn't you?" he said, and I thought, here it comes, here it is, and all I said was, "Yes. Yes, I did." I wondered which question he'd ask then—Why did you leave? Why did you get fired? What happened?—but he didn't ask any of them; instead what he said was, "Do you like it better on *Today?*"

"I like it very much," I said, "and I'm very grateful to Tom Redford for giving me the chance," even as I was saying it, thinking, "What the hell are you sounding so defensive on Redford's behalf for—who told you you had to preach *that* gospel?"

"Oh, Redford," he said. "He's a fool. But I like him."

"Why a fool?" I asked, mentally kicking myself for this save-Redford routine I seemed to find myself compulsively putting forth.

"He's a con man," Wolffe said, in quiet and absolute dismissal. The way he said it you could almost see Redford silently disappearing in one puff of smoke.

"Shall I order for both of us?" Wolffe said, and I smiled, agreeing, thinking to myself: what old movie did you see that in when you were a kid and now you've finally gotten around to having a chance to say it—chalk up one more giant step for little Davey Wolffe . . . the skinny-kid, marble-collecting fencer.

I was so fascinated by the studied quiet pitch of his voice as he gave his order to the waiter that I almost missed all of the carefully "French" things he was saying, except that you couldn't help noticing how polished it all was, right down to the final, "*Excellent—merci,*" and the sort of invisible flourish with which he dismissed the waiter. Grudgingly, I had to admit that it was a flawless performance. "*A* for elegant," I said to myself, "also ass-arrogant," only to have him turn to me just as the waiter got out of earshot and say, "You wouldn't believe it, but the first time I saw a finger bowl I thought it was soup." He grinned, remembering the incident, and I thought, "*Who* wouldn't believe it?" but it still wasn't *that* easy to dismiss the fact that he had said it about himself—and that he had grinned.

"Have you ever met Arthur Jefferson?" he asked me.

"You mean . . . the chairman of the Board?"

"Yes," he said, and grinned. "I call him the Queen of Hearts—you know, like *Alice in Wonderland*—" and in a deep, Queen-of-Hearts voice he pronounced, "Off with their heads—off with their heads—"

I couldn't help laughing; it was a pretty funny performance, especially since, for a long time now, I'd equated a lot of what was happening with the madness of *Alice in Wonderland.*

"You should see the Board meetings," Wolffe said. "Or better still, what he refers to as one of his informal conferences." And in his Queen-of-Hearts voice Wolffe said, " 'Just a little pleasant chance for us to chat among ourselves'—except that he doesn't permit any booze or ashtrays or even coffee. It makes for quite a relaxed gathering." He grinned, "Once when Tom Redford was there he took out a cigarette. I guess nobody had told him. It wasn't more than two seconds before one if the secretaries hurried up to him with this little flowered dish. I think he thought she meant it to be an ashtray until she sort of whispered, 'It's to put it out, sir. For Mr. Jefferson.' I can still see Redford stomping out the cigarette while the secretary held that little dish for him. I can still hear him saying to her, 'Hell, m'am, why didn't you say so? You looked so put out I thought maybe I'd neglected to button myself up properly.' "

Wolffe laughed, remembering Redford.

"Damn fool," he said. And then shrugged and grinned, "Except he does have swagger. *Brio*, the Italians call it." And I wondered where he'd picked that up.

Suddenly, almost without a break, he said, "You should be home."

"Home?"

"Writing," he said. "Full time. Every day."

"Oh," I said, thinking: how in hell do you know that—what am I to you anyway? What crooked mess is this whole thing all about and how soon, if ever, am I going to find out where it all leads to?

"Are you going to give me the book?" he asked.

"Oh—well—I don't even know if I have one"—and then, hearing the ridiculousness of that, "just my mother's copy, of course, and that's got writing all over it, and anyway it was a long time ago."

"I'll get Brentano's to order it for me. How long ago was it published?" he asked, and I smiled and hesitated for a minute and said, "Oh, it must be at least four or five years," knowing it was seven, knowing that even *I* had exhausted the possibilities in the "Books, Located . . . Try Us" section of *The Saturday Review*.

"It means your ancestors were from Spain or Portugal . . ."

"I beg your pardon?"

"Sephardic Jews," he said. "What I am on my mother's side."

"I know," I said. "You told me."

"Oh. That's right, I did, didn't I?" And then he said, "I thought of writing a book myself. I just don't have the time." I smiled, as cordial as an iceberg, and said, "Oh, is that what it is—a matter of time?"

"Of course," he said, rearranging the silverware as the waiter arrived with our lunch. "Anyway, I know a lot about people —I mean, I understand them." It was almost as though he stopped, waiting for me to contradict him or argue or something. At any rate, when I didn't, he said, "At least more than most people—and that's all a writer really has to know . . ."

For God's sake, he's trying to *provoke* me, I thought, the word coming out of nowhere, except that that was what it seemed like. Almost as though he wanted someone to disagree with him, to think differently—to *argue*. In that same second he said, grinning: "You're *sure* you're not afraid of me?" and I thought: Whee! I *know* you . . . I *know you* . . .

I didn't even bother to answer this time, and for about two minutes we ate in silence, and I thought: *What scares him is that everybody is scared of him.*

"Who do you live with?" he said, breaking into the clue I had found, so that I thought: I'll think about that later—that's really all of it. But not yet knowing exactly what I meant.

"I live with my mother."

"Doesn't that make it rough for you?"

"What do you mean?"

"Well . . . you resent it—at least sometimes," he said. Automatically, and honestly—as far as I could tell—I started into my speech, the one that goes, "I guess in almost any other case that would be true. It's hard to explain, but with my mother it's completely *different* . . ." It was a speech I had made many times before. Because it's true. My mother is a kind of a nut, I guess, and yet I couldn't ever for a minute say that she "got in my way" about anything. Not even Steve.

All of a sudden the need to hurt Dave Wolffe swept over me, swift and unexplained. Unexplained because I guess I didn't like people pulling covers off things that had been neatly and comfortably settled over for years. What business of his was it where I lived, or who I lived with, or whether I was married, or lonely, or—

"Shouldn't you be getting back to work?" I said coolly. "You must have a very full schedule."

From that point on, getting our coffee and the check and everything else went pretty fast, and if I still felt like Alice in Wonderland, at least I felt that I had put one over on *this* particular Mad Hatter. The feeling lasted all of about five minutes. As he stood behind me helping me into my coat, he said, smiling wryly, "You realize that none of this would have happened if you hadn't stared at me in that meeting last week."

I turned around almost furiously, but he had signaled the doorman, who had signaled for a cab, and now he stood holding the door open for me, cutting off anything I might have said.

As I got into the taxi, I realized that he wasn't coming back with me. "I'll walk," he said. "I've got three editors waiting in my office and it might be fun to give them a chance to think for a while." He grinned from ear to ear as he slammed the door shut for me and thrust some money at the driver, so that there was no chance to talk, no chance to say what do you mean, what's this all about anyway, no chance to challenge; no chance to do a single goddamn thing.

"Where to, lady?" The driver looked over his shoulder. He had a little brown cap, and the sign said his name was Samuel Morris and what did he mean, "If you hadn't stared at me" and—"God knows," I said to myself.

Then I gave Sam the address.

eight

IT STARTED TO RAIN AFTER I GOT BACK TO the office. "God's crying," I thought without thinking. And then I thought, *damn it,* when are you going to stop reacting the same way you did when you were four years old?

Right then Jack Sheehan came back from lunch. Hat dripping rain. Hair dripping rain. *Glasses* dripping rain. Stoned.

I decided to ignore him.

"Damn professional southerner . . ."

I said, "Excuse me?" But it didn't matter. Jack wasn't really talking to me and the subject of his monologue immediately became perfectly clear. "Big Daddy Tom—Goddamn it, he's been up North for twenty years, why does he have to talk as though he were straight off the plantation? Sons of the old South. Redford from Mississippi, me from Arkansas . . . Maybe that's what I should have taught myself to do, maybe I'd have ended up being publisher instead of a promotion writer." Half laughing now, a little hysterical, a lot drunk, "Hey, that's it, maybe that's all I have to do, just practice up a little," doing Redford, wild, broad, so southern-thick you could hardly understand him, yet he was it, he was real down-South, never-left-the-darkies Redford, thick, wild, familiar, "How about that, bud-dy? . . . how's that, ol' bud-dy . . . what do you think of me now, ol' *bud*-dy boy?"

It was so good I almost got carried away by the show business, so good that I almost forgot for a second how furious I was at Sheehan for being such a dope, how I hated him for it—getting drunk, *coming back* drunk, risking his job, making *me* feel rotten. Maybe-knowing suddenly-maybe almost all.

"*A sweet kid, huh—wasn't I a sweet kid?*" That was the chicken bone, half-up, half-down your throat. Stuck. Wedged. Dave Wolffe had said that. Grinning. About himself. Dave Wolffe. Hawk. Vulture. Vulgarian. Except maybe more honest than I was. Like maybe one hundred percent.

I don't know where my little one-woman soul-show would have taken me, because my phone rang right then. It was Susan Planker, Mark Post's secretary, asking if I would come into his office. I don't think Sheehan even noticed my going.

As I went up to Post's doorway I saw him sitting up straight and serious behind the carefully arranged desk. He was like someone in a cartoon with a balloon leading into his head saying: "Man: Thinking." He looked up quickly, stood up, and said, "Come in, Paula. Sit down. How are you?"

What flashed through my mind immediately was: "My God, Paula, how sick you are. One of the few nice ones in the whole group, and he's the one you don't cotton to at all."

"It's good to see you," he said. "I've wanted to have a chance to talk with you ever since you came to *Today*. Unfortunately, I seem to spend a great portion of my day dealing with the *past*. Listening, and listening and listening. Even when I was down in Atlanta I always respected highly the élite corps of Kimberly, but what I underestimated was just how *articulate* the *Today* salesman could be."

I thought, It sounds as though he's dictating a memo. He was good-looking, though, in a thin-nosed, kind of snotty way. Definitely not my type—whatever the hell that might mean. "Anyway," he said, smiling, "what I really wanted was to thank you for this." He held up a copy of the letter I had sent to him and Redford. "I appreciate it very much."

I managed to say, "Oh, well—I guess I just felt like writing *something* . . ."

"Well, I think it's fine."

Suddenly, I almost saw it from Redford's point of view. Maybe I just wasn't used to people being nice any more than he was. I started to get up out of my chair, feeling kind of foolish, except that as I was doing so Post started talking again. "Actually," he said, "I think as soon as the sales staff realizes how much I trust them—their own judgment—things will be much simpler. That's what I've been trying to tell them. If there's an immature child on the staff who can't handle his emotions and his work as an adult, I don't know about him. Tom Redford and I respect their dignity, that's the point I've been trying to get across to every one of them that comes in here. That there's no reason why this can't be as proud a group as any in the business community. The way I feel, every *Today* publishing representative . . ." (that stopped me for a moment, until I realized that it was Post's way of saying "salesman") ". . . every publisher's representative has a Canfield-given right to swagger a bit."

I almost felt as though I was supposed to applaud. Not that I didn't agree with him. It was just——

"You were a good friend of his, weren't you?" I asked. "Mr. Canfield, I mean?" I could almost hear Redford: "*You just don't know your place, is all* . . ."

"Jim Canfield was the greatest ally the sales department of *Today* ever had."

It occurred to me that that didn't exactly answer the question I had asked, but anyway . . .

"I don't know Ted Monger very well," Post said. "What I've seen of him I like. What does get my sweat up is the way some of Canfield's old staff are reacting like spoiled, unmanly, petulant *children*. Sometimes I have the strongest impulse to just let them strangle in their own bile."

"Orin Kreedel?" I said very quietly . . .

"All of them, mostly, except he's the worst. I love that self-centered, crazy bastard. But he's impossible."

"Was he supposed to be editor?"

"He would have been a good editor," Post said, "I suppose any of a half-dozen people would have made good editors."

I choked a little on my spit, but I managed to keep still.

"Do you know what he said to me the other day—Kreedel? I asked him to sum up exactly why he finds the present editorial structure at *Today* 'impossible' as he calls it, and he said, 'Because Monger doesn't have taste.' It sounded silly, and I told him, so." He was talking fast now. "Taste," he said mockingly. "Why, Harold Ross didn't understand half the stuff he put in *The New Yorker*. Naturally Monger doesn't have the *religion* Kreedel has about *Today*. How could he? But give him some time. I'm not saying things will be better six months from now, but what they've got to do is open their minds a little bit. Of all the people I know, Orin Kreedel would be the first to object if someone said, 'I think *Today* is perfect.' Who knows? Monger might be just the guy to help improve it—maybe not. Anyway, a little fresh blood never hurt."

I had a spooky feeling he was saying it to me exactly the way he had said it—or even written it—before. It was all too pat, too rehearsed. Also it was a little too much like: If I say it again maybe *I'll* believe it a little more this time.

"Look," he said, "neither Ted Monger, nor anybody else, is going to ruin this magazine. If Kreedel—if any of them—love it as much as they profess, they won't permit it. Neither will they permit themselves to indulge in peevish colic deadline-ultimative gestures. Either stay and battle it out, or leave quietly. That's what I told him."

After a moment he gave me a sickly smile. "It's different than I thought it would be when I came here from Atlanta," he said. "Would you believe I used to sight-read music for a hobby? Now this desk and I are practically married to each other" ("*too busy even to care whether he gets laid or not*"). "I'm beginning to find my way around, though," he announced, suddenly being very positive. "I went to the Philharmonic on Saturday, and I've promised myself to get to the Frick this weekend even if it means Kimberly folds down." He smiled broadly to show me how funny he was being.

"Well," I said, standing up, "thank you for . . . thank you."

"Thank *you,*" he said. "I didn't want you to get the impression that the entire executive staff you work for only reads mystery books."

"Oh—" I said, "well—no, no, I didn't."

"As a matter of fact," he said, standing up to let me go, "I still get in a few licks on the saxophone every once in a while."

"Oh—I didn't know you were a musician too."

"Oh, I dabble with the sax, and the trombone—the oboe a little. Sometimes I even break out the drums. As a matter of fact," he said, coming around the desk, guiding me in a noncommittal, executive-friendly way to the door, "the only thing I do not understand, and therefore do not play, is company politics. ("*The only trouble with Mark Post is . . . he's never been to the fight.*") Post smiled warmly. "I *am*, however, a card-carrying member of the Museum of Modern Art. Or MOMA as they call her." I gave what I hoped was the appropriate smile.

And suddenly I couldn't help wondering where—when the chips were down—Mark Post would choose to throw his. Would he side with Rusk and company? Or Moshier and his boys? The fact was that he didn't really seem to belong on either side. But one day Post would have to choose. And I wouldn't give two cents on which way it would go.

We were at the door now, and he said, "Thank you, Paula. Again. And I mean it." The spookiest part of it being that I was sure that he did.

When I got back to my office Jack Sheehan was standing in the middle of the room, looking as though he had either just finished or was about to start a speech. My entrance, obviously, was his cue to begin.

"I'm sorry, Paula. I didn't mean to make such a damnfool ass out of myself before."

"It's this place," I said. "I know all jobs are probably full of kooks, but Kimberly has *got* to be the leader."

"What do you expect with old *bud*-dy 'Big Bob' Moshier running the show?" he said. "Old make-it-while-you-can Moshier. You think *he* cares about *magazines?* Another star in his slightly tilted crown, that's all bringing Kimberly out of the red'll mean for him. Assuming he can do it."

"Is that so bad?"

"I don't know," he said, suddenly very quiet. "Maybe not."

I said, "At least Mark Post seems relatively sane. He's really pretty bright, I guess. No wonder Redford's always talking about how smart he is."

Jack laughed. I had never heard him laugh that loudly before. "Ol' *Bud*-dy Big Daddy Tom, you mean . . . praising old work-a-day Mark? This is Mark. Mark has a book. Mark has a nose. Mark keeps his nose in the book. This is Tom. Tom has a nose. Tom sticks his nose into Mark's book. This is a fight . . ."

"What do you mean? Redford's always praising Post. Telling you how smart he is, how he's always working. The thing is that in their nutty way they do make a perfect combination—Post always working, and Redford—Redford being the big showy up-front character that handles the smooth, high-level conning. What could make a better publishing team?"

"Nothing," Jack said. "It's perfect. At least on paper."

He stopped talking then and this time I didn't try to draw him out. Just then I realized how tired I was. It was getting to be a pretty long, confusing day. Suddenly the idea of going home and going right to bed seemed the most beautiful thing in the world I could think of.

The phone rang.

"We didn't really have a chance to finish our conversation." It was Dave Wolffe.

"Oh . . . oh, well, thank you for lunch, I mean—I enjoyed it very much."

"Could we finish it sometime? Maybe at dinner."

"Oh—oh, well, that would be fine."

"What about tonight?"

All I kept remembering was to keep it cool. Unfortunately, that took the form of my saying, "Well, yes—yes, that would be fine." Thinking, bye-bye bed . . . bye-bye going right home and right to sleep, bye-bye everything nice and quiet and——

"I'll send my car for you—eight o'clock. Is that all right?"

"No, no, that isn't necessary."

"Are you sure?"

"Yes, yes, I'm sure." Of what, in God's name, I wondered.

"All right then. I'll meet you at the Plaza. In the lobby. Eight o'clock. Don't be late," he said. "I hate waiting." I was much less than half certain that he was joking. He hung up.

All the rest of the afternoon I kept thinking: he'll call it off, he'll call back, the phone will ring, something will come up. Instead what caught my attention was Mark Post's sudden hurrying down the hallway past my office. It occurred to me that he walked like the rabbit late for his appointment in *Alice in Wonderland*. ("Dear me, dear me, what will the Queen say . . .") It also occurred to me that I had never seen him out of his office before. It was another "first." I wondered what would happen next, since obviously—the way things were going—*something* had to.

What happened next was my phone ringing. It was Jane Perry.

"Paula?" she said.

"Yes—Jane? is this Jane?" I had never heard her sound upset before.

"Paula—" It was as though all of a sudden she couldn't remember exactly why she had called me.

"Paula . . . I guess . . . well, everybody's going to know in a little while anyway. Paula, Orin Kreedel's been fired."

"What?"

"Just this afternoon. Evidently some of the other editors are planning to walk out because of it . . . well . . . I don't know . . . I just thought . . . Mr. Redford isn't here . . . I just thought he might want you to know . . ."

"Yes—well, thanks, Jane—thanks for telling me."

I hung up. My brain seemed to have been short-circuited. Only one thing came through. Now I wouldn't have to go tonight. Now Wolffe wouldn't be able to go. He'd have other things to do. Now I could go home and go to bed and go to sleep and——

The phone rang. It was Dave Wolffe.

"You remembered it was the Plaza?" he said.

The blood in my fingers started to buzz.

"Yes," I said. "The Plaza."

"Eight o'clock," he said. "Don't forget now."

For a moment after I hung up I thought of every dirty word I ever knew, but after you've gone through saying them to yourself, it's amazing how few there really are. And how even *mild* they seem after you say them over and over again a few times.

But there was one shocking thing. One phrase so dirty and filthy and shocking it was like the first time I had ever seen "Fuck you" printed on a brick wall. It took my breath away the same way that had done—just raw, hard, brutal filth. And it went: "They fired Orin Kreedel."

I was going to call Wolffe back. I wasn't even going to speak to him. I was going to cop out; I was just going to leave a message with his secretary, old, birdlike Eileen. I was just going to say I couldn't go, I didn't feel well, my uncle had died, a filling had come out, my mother was sick, I——

"Just pray you end up knowing who you are," Jim Canfield had said. "God help you if you ever forget that. So—what are you going to do?"

The hell with you, I thought, the hell with copping out. It was too easy that way. I'd go. Who knows, one way or the other—one minute or another—stupid, stubborn, but maybe knowing more than I knew I knew—I'd go, and in some way, or other—maybe one way or another—I might even get in a blow or two for—for something or other.

"Fuck you," I thought, "fuck all of you," but it seemed nothing—kid stuff, mild, silly. It was nothing—pure nothing at all, compared to the really filthy truth of what had happened.

nine

THREE HOURS LATER I FACED THE FACT that here I was in the lobby of the Plaza Hotel ("*Don't call Steve*") waiting to have dinner with Dave Wolffe, who I hadn't even wanted to have *lunch* with, and Orin Kreedel had been fired, and my hands felt sweaty. And it had all happened in the same day. In addition to which, Wolffe was now ten minutes late. I covered another yawn, my fourth in the past five minutes. It was, as always, my infallible indication of resistance. A *grande dame* in a mauve tulle hat swiveled around on the red velvet couch to stare at me. "Drop dead, Lady," I thought, and smiled my sweetest Catholic-school smile at her.

I had a crazy impulse to break all those fancy little showcases in the lobby with their snotty little hands-off displays of Je Reviens and Intime and those real alligator handbags by Lederer that I've promised myself I'm going to have one of before I die, or burst . . .

And then I had to admit that in addition to the nutty feeling I was having because of everything that had happened at Kimberly that day, what I was feeling most strongly that very second was sorry. For me. My little-match-girl-out-in-the-snow feeling. *Poor thing . . .*

I don't know how I missed seeing him come in. All I know is that someone said coolly, "Hello. You've got your hair different. I like it."

Well, thanks a heap, I thought. I took my time standing up, waiting for his excuse for being fifteen minutes late. It never came. I started to move into the lobby when he said, "Let's go someplace else. All right?" It was a purely academic question, since he had taken my hand and started to lead me out of the hotel. It was bitter cold in the street, and I grabbed my coat around my neck with one hand, amazed to find that he still had hold of the other.

"Do you mind walking?" he said. "It's only about a block."

My teeth were chattering, and I didn't know what to think—much less say. By now we were sort of running along the street, barely missing a hansom with a frozen old horse trudging out of the park.

For a split second, the pain was the smooth stab of an icicle . . . It was like something happening that I knew—something warm—old—the deepest part of me. Something I—*recognized.* My hand in his—running . . . Except there was no time. Not now. *Later*—I would have to think of it after—later—whatever it was . . .

Out of breath and half frozen, even though it had only been about a block, we finally came to a little restaurant called Chez Vito. I had never been there or even heard about it.

"I think you'll like it here," he said.

We went into this dark, warm place. There was a woman singing some kind of Hungarian song and, though the headwaiter smiled graciously, the waiter did not seat us until she had finished. It all looked very red and gold and expensive.

I was glad for the two minutes we had to wait, because now—with my hand released—I was able to think clearly again.

The first thing I forced myself to remember was who I was with and that he represented practically everything I hated, and that, besides that, he was cocky and conceited and liked Brad

Rusk and thought anybody could write a book if he just had the time.

The second thing I made myself remember was that he didn't look like anyone I could ever imagine myself being at all interested in. I made a whole catalogue of the points: of how he was long and skinny (*I keep in shape by fencing every weekend.*) and had thick lips and a broad nose and how his brows were kind of heavy with bone, and how his skin had this kind of cold-touching look to it (even if his hand had been warm).

By that time we were seated at the table, and the first thing he said was, "You know, you look a little like me." He said it absolutely straight-faced. Immediately afterward he said, "You look a little like Nina Foch, too . . . sort of femme fatale." Then he said, "They're going to have your book for me tomorrow. They had to send to someplace in Boston. By the way, did you know it was out of print?"

I smiled and didn't say anything.

"When are you going to publish another one?" he asked, and for a second all I could think of was, "Not *you,* too." I was half tempted to say, "When I stop getting involved with doing nutty things like being here."

Instead I said, "I'm working on one right now," hoping that no one would ask me to explain how "right now" meant on and off for the past seven years.

"Do you have any brothers or sisters?"

"A brother. He's married."

"What's his name?"

"Frank."

"What does your brother do?"

"He's an accountant."

"Where?"

"At *Newsweek.*"

"Do they live near you?"

"A few blocks away."

"Do you like his wife?"

"Yes." If I showed any outrage or surprise at any of this, it obviously made no difference at all.

The questioning went right on—as personal, as curious, as cross-examining as before.

"She's a very nice girl."

And then, without even a breath in between, "Why did you break off with him?"

Pretending that I didn't understand, I drew back the breath that had been knocked out of me and said, "Who?"

"The man you used to go with."

I wanted to say, What business is it of yours? I said, "I had to."

"What does that mean?"

"Just that. I *had* to."

"Did you know him long?"

"Yes. No—I don't know. About seven years, I guess."

"And you really loved him?"

Drop dead, I thought. Except that I had started. I had never said any of this before. Not even to myself. But why now? And why to him?

"Yes. I don't know. He was wild and noisy and larger than life and he made everybody else seem like a shadow. Except, at the end, we couldn't be together for ten minutes without clawing each other."

"Why?" He was as implacable as the devil; there was no getting away.

Except this time I didn't answer him.

"When was the last time you were together?"

"Two months ago."

"And he's never tried to see you since then?"

"No."

"And you've never tried to see *him?*"

"I don't let myself. I *remind* myself. I've got it written on my calendar—" I stopped. I had never told anybody about that before. *Why now? And why him?*

"What was his name?" He asked very quietly.

And I said it out loud. "Steve."

And then, when there was nothing that could be said, he dropped it, and grinned. "I wonder what I would have done if you'd have said no about tonight . . ."

"Excuse me?"

He didn't repeat it. He just sat there, and the thing that threw me most wasn't his surprising irony or gentleness, but the fact that he permitted himself to be seen being vulnerable.

I started the speech going over again in my head: "He does not know anything about magazines . . . he's got the impression he can have anything he wants . . . he thinks Brad Rusk is a great editor . . . he's *trouble*."

And right there I remembered something I had wanted to add to the list.

As though reading my mind—*again*—he said, "I didn't want Orin Kreedel to leave. There wasn't any choice."

I didn't say anything, but thought, "Bullshittt . . ." as they say in the comics. Which had been a favorite expression of Steve's. (*Don't call Steve.*)

"The others are leaving, too. It won't all be announced until tomorrow."

"The others?"

"The other four editors besides Kreedel who worked for Canfield."

"They're *fired?*"

"It's all sort of fuzzy," he said. "The five of them were a kind of package. It was all of them or none."

"And you let them go?"

"They couldn't work with Ted Monger," he said. "They made his life hell."

I didn't say anything. For a kid who'd talked a lot most of her life, I seemed to have slipped into an interesting period of discreet silence. Otherwise known as cowardice.

"They started petty arguments with him. They wouldn't cooperate. They said he didn't have enough 'taste' to be editor . . ."

"Let me order," he said. "Would that be all right?"

It was with that sentence that I saw, for the first time, what was to become a very familiar gesture. And a very dangerous one. The way he looked at you, asking you to say yes, sure you would say no—. . . maybe it was that his open vulnerability was such a shock after the cockiness and arrogance that seemed such a natural part of him. Maybe it was that I hadn't been prepared to have to keep reminding myself every two minutes that he was the villain. Whatever it was, the way he had asked the question didn't leave much room for any answer except yes.

"I made a speech to the editorial group at *Woman's World*

"You look like a villain when you frown." I don't know why I said it. I hadn't intended to make light conversation, certainly I hadn't meant to say anything funny or personal.

"Thanks a lot," he said.

"I mean you look as though you were practicing to tell everybody they were fired, or that the stock market fell or something."

"Please—not so much flattery all at once." He grinned.

The cab came and we got into it and he told the driver to go to the St. Regis.

It was a very short ride, but neither of us said a word the entire way. When we got there he took hold of my arm just above the wrist and led me down some steps to a place called the Maisonette. He checked our coats and then very coolly signaled for the headwaiter, who led us to a banquette near a small dance floor, where a lovely blonde girl in a high-necked, plain black dress danced in a stunningly perfect way that made the rest of the people on the floor all seem like a kind of backdrop for her. I wished I were home in bed. Under the bed.

Quietly, Wolffe stood up from his place next to me, walked around the other side of the table, and sat down on my left, his back completely to the dance floor—*and* the blonde girl.

"Hey, you're supposed to pay attention to *me*. I brought you."

That was when I started concentrating again on how terrible-looking he was, except I didn't get very far, because he was saying, "Did I tell you I heard Du Pré play last night?"

"No, you didn't."

"At Carnegie Hall—she's a fantastic cellist," he said. "I go there a lot by myself."

I didn't say anything. In the back of my mind I said, "Why? Are you thinking of buying out the place?" It was a new defense I had discovered for myself, concentrating only on those remarks of his that seemed to be what I thought of as "in character" (like, "Anyone can write a book. It's just a matter of having the time").

"I guess the music I really love the most is Bloch's 'Schelomo,'" he said. "Do you know it?"

"No, no, I don't." I decided to play this one straight for the moment. There seemed to be less risk involved that way.

"*Schelomo* means *Solomon*," he said. "It's for the cello. The way I feel about it is—well—" For the first time he hesitated for a word; it seemed important to him that he say it right. "It's one of—well, I'd say the saddest, most despairing things I've ever heard. It's like—well, the way you feel the moment you realize you've just been completely happy and now you're not any-more." He was quiet for a minute, but you could tell there was more he wanted to say.

"It's mysterious, too, in a way—and I guess—well, disillu-sioned. But—mostly—it's about things that got lost."

After saying this he went absolutely still for a minute, and then, when he spoke again, he had changed back to his usual sort of flip half-brashness. All the vulnerability safely hidden, he said matter-of-factly, "I'll take you to hear it sometime."

It was as though I had to say something that would keep me from drowning in this thing that was happening to me, this feel-ing of being in grave danger. I said, "I don't really like music very much."

"Oh, but you will," he said. "I mean—you—you'd have to . . ." And suddenly, not knowing why, I felt ashamed, and small.

"After a while you get to associate music with people. Dif-ferent pieces, that is."

He ordered us another drink. I tried to concentrate on how much I wanted to be home.

"The first time I heard the 'Schelomo' my cousin Moisie played it at my uncle's house. My uncle was dead. He'd been run over on Third Avenue and there hadn't been any man in the family to identify his body. Moisie was only nine—and my own father was dead—so I was the one to go to the morgue to make the identification. I couldn't believe it was my uncle. It was as though the picture of him lying dead in the street had taken hold of me—that this man lying on the table in the white sheet—over his face and then the way they flipped it off for a minute—as though I was looking for someone else. I told them yes—and then I went back to my aunt's. And that was when my cousin played the 'Schelomo'—except he didn't have a cello—he played it on a violin—and we stayed up and talked and my mother came and there was a lot of food and we ate. But all the while I kept thinking, Where is my uncle really? and my cousin kept playing the music . . . I guess I was about fifteen . . ."

I thought, "I should have said no on the phone—I should have called back—I should have left a message."

His eyes were cold, like the eyes of someone who has been stolen from, but just behind them there was something like a terrible warmth—something I did not want to see.

"We all see people lying in the street sometime in our life," he said. "Just last week a man had a stroke. But for me it's always Uncle Herbie. Come on," he said, suddenly efficient, cold, "Let's dance."

"But I don't——"

"Come on," he said. "You'll be great." He had taken hold of my hand; there was really no choice at all.

This was the smallest dance floor I had ever seen. And there was a mirror down one whole side of the wall. It was like being on display on the top of a table in the middle of a department store window. I thought, "Queen of the most holy rosary, help me," which is what usually comes into my head when I'm completely *in extremis.*

"Relax," he said, "I know what I'm doing." I was sure he did, but at the moment I couldn't see how that was going to be of any great help to *me*. The amazing part of it was that, after a minute, I realized that I was dancing with practically none of the trauma that it usually involved. They were playing "All or Nothing at All."

I said, "I told you I couldn't——"

"Shhhh," he said, "listen. It's a great old song."

So I "shhhhed" and listened. The piano player began to sing. It wasn't bad. I found myself—if not enjoying what I was doing—then at least enjoying the fact that I wasn't terrified by what I was doing. When it was over, I think I practically *preened* a little on the way back to the table. I actually felt as though maybe the next time somebody asked me to dance I might not want to slip behind the woodwork. "See?" he said. "You were great. I told you."

We had no sooner gotten back to the table when he finished off his drink in one swallow, signaling for another. With almost old-movie-style disdain he said, "I drink quarts of it—it never affects me." Somehow I doubted that, but I respected the authority with which he said it. "I'm flying the company plane to Dayton next week," he said. "Will you come with me?"

"Oh, do you fly?" I fully realized that it was only a temporarily effective delaying measure, but I was more than grateful for even that.

He laughed. "You think I'm like that phony Bob Moshier, with his personal little pilot? Hey, didn't I ever tell you about Bennington?"

He smiled like a four-year-old kid about to tell you how he won the skating prize. All you could do was smile too, and look eager to hear. And try to convince yourself that you aren't.

"It was when I just came to Kimberly," he said. "I had an idea for an art insert in *The Review*. This little museum up in Bennington puts out this booklet on American embroidery. All they do is sell it through the little museum up there. I figured if I could buy out their whole supply and bind it right into the magazine it'd be worth at least a quarter of a million dollars in extra hobby-type advertising in that issue. So I arranged to fly to Vermont. But all I could schedule there was forty minutes, because I had to be back in New York for a meeting. Well, when we got into Vermont I figured, with all the Avis and Hertz cards, I'd just rent a car, except there weren't any cars.

"Then I remembered that as we were coming in I'd seen a Pace Flying School. I hadn't flown a plane for five years," he laughed, "and I hadn't been very good then. Anyway, I get a fast hitch over to the field. There's no maps, no directions—nothing. The guy points to the left and says, 'You fly right over that hill over there—you see it?' And that's what you do. And when you get over the hill there is a landing field, except it isn't a landing field, it's a strip. And in the meantime you've been trying to raise the field on the radio, but you can't raise anybody. So what you do is pray that nobody is putting down at the same time you are."

He took a deep drink from the glass, still smiling broadly. "And so I landed. And there's a guy playing with his plane on the ground. And I run over and borrow his car and drive it in to Bennington and close the deal in ten minutes. And then drive back. *And* fly the plane back. And make a quarter of a million dollars."

At the same time I thought, it's crazy. Except it was beautiful too.

"I'm going to have Kimberly buy me my own plane," he said. "Did I tell you I'm teaching Brad Rusk how to ride?"

"Horses?" I said, somehow not at all sure he didn't mean a rocket, or a tiger, or——

"He's scared," he said, laughing, but fondly. "I'm going to start a Kimberly stable," he said. "There's this old gangster restaurant owner on Fifty-sixth Street who's invited me to use his lodge in New Hampshire."

For a moment I thought of Jim Canfield—"They'll tell you all about being invited to swim in the White House pool. And isn't it a shame for the image that Johnson has so much flab."

"Didn't Brad Rusk mind your just telling him what to bind into a magazine?" It wasn't much, but it was the most I could think of at the moment.

"Bind in . . . ?"

"The museum thing."

"Oh," he said, "I wouldn't have done it without Brad's wanting it. He thought it was a great idea as soon as I told him about it." And then, after a moment, "Brad's a great editor."

I looked at my watch. It was like checking your gun, if you had a gun. Just to see it was there. Maybe not using it yet, not this very second, but checking to make sure it was there. "You're sure you're not afraid of me?" And I laughed instantly, hilarious even at the idea of it, although interestingly enough I did not answer the question. Instead I thought, "When are you really going to tell him what you think of Rusk?" And then for a moment, strongly, the idea came back that *now* he would ask about Kimberly—what was wrong with it—about *Woman's World*—about work—about what he wanted to know.

"Do you know what I feel like doing right now?"

"What?" I said.

"Kissing you."

"No." I shook my head.

He leaned over the table and his lips touched my mouth

gently, and, bewildered, I made no effort at all to resist his kiss. The waiter came back.

"Another round, sir?"

"Yes."

Then, just as the waiter left, he said, half-mockingly, half-vulnerably, "Aren't you afraid you might fall in love with me?"

I swallowed and smiled. Coolly, I hoped. But not too sure.

"It could happen, you know," he said, smiling. "Maybe not right away. But in time you might fall very much in love with me. And then after a while, without ever saying why, I might just decide not to see you again, and then you'd start trying to find out why and you'd keep calling me and leaving messages, only I'd never answer any of your calls."

There was a certain broadness with which he said this, a certain deliberate broadness, almost like telling a very obvious joke —but not *exactly* a joke—and I played along with it, laughing or whatever the sound was I managed to make, at his shyness, his really terrible false modesty, ". . . really, you ought to do something about such shyness . . ." Except that, underneath it all, the compulsive chatter and stiff, necessary smile, I was cold. "It's pretty late," I said, "I'd better go home."

"I'll call the car," he said, and—before I could object—"I'll need it to drive me home anyway."

I waited while he was gone. I didn't know what I was thinking. It was all wild, and impossible, and if I had any sense . . .

"Let's have a Drambuie while we wait," he said, back before I could really think anything.

"Where do you live?" Just to say something—because I knew.

And he said, "Forest Hills," and then quickly changed the subject.

"Did I tell you about this recurring dream I have?" he asked me, and I said, "No," no longer even trying to tell him that I wasn't someone he *told* things to, that he must have me mixed up with somebody else, that we were practically strangers—all I was was somebody who wrote promotion for *Today* magazine, and the only reason we were even together this once was . . . was . . . I never did find the end of the sentence, because he said, "I'm a king. In this dream I'm a king. And I'm sitting in this empty throne room, with the long robe and the thing on my head, and I'm crying. And my friend comes in and says, 'Why are you crying, Dave?' and I say 'Because I'm king and I don't believe in God and my father is dead and that means there's nobody who knows any more than I do.' "

Neither of us said anything. "Let me sleep on your couch tonight."

"What?"

"Why not?" he said, "there wouldn't be anything wrong with my sleeping on your couch. You could sleep with your mother this once, couldn't you?"

"My mother'd probably end up on the floor. I kick a lot," I said—and then, quickly, just to keep talking, "maybe it's because I have this recurring dream of my own."

"Tell it to me," he said.

God knows a writer should at least be able to make up a *dream.* But leave it to old Paula to play it straight—especially at the wrong times. "There's always this big, black panther," I said. "Like last week. I dreamed that I was coming out of this movie house. The sun was very bright. And for a whole block the street was entirely blocked off. I could see the people being held back by the police. I was the only one out in the open, all by myself, with the sun shining. And this big black panther came right up to me—slowly—walking around me, smelling me. And I was absolutely petrified. All I knew was that I had never been that afraid in my life. Except that's the way I always feel in the dream."

He stood up.

"Where are you going?"

"I'm going to tell them not to send the car. I'm going home with you."

"No, you can't." But he was gone already.

By the time he returned, annoyed, and said flatly, "Damn it, the car's on its way," I thought, thank God—thank somebody. Thank whoever it was that was answering the prayer I hadn't had sense enough to say.

"We'll just send it back," he said.

"No," I said. "Anyway, I've got to go home now."

"What are you so afraid of?" he said flatly.

After that we didn't say anything, until he suddenly turned and said: "Do you realize you said petrified?"

I looked at him.

"Petrified?"

"Outside the movie," he said, "you said, 'I was petrified.' As though you couldn't run." He grinned: "That big black panther is pretty important to you—you know that, don't you?"

Now it was my turn to grin. Because that—of all the insane things I had heard that night—was the nuttiest of them all.

"Sure," I laughed. "And so are elephants and rhinoceroses and purple-and-yellow butterflies."

"You know what I'd like to do to you right now?" he said very quietly.

It was like a sponge being squeezed tight down in my belly . . .

And then, there was the tall, muslin-faced headwaiter, or whoever he was, saying very quietly, "Mr. Wolffe—your car is here now . . ."

At the sidewalk the 7Z chauffeur couldn't have been more solicitous if he'd been for real. He ran from the car to the doorway and then back to the car again to hold *that* door . . . And the discreet thing with the glass partition closed tightly just as soon as Dave Wolffe had given him my address made me feel as though just riding home was being in the middle of the most complete Roman orgy I had ever imagined.

Except that the atmosphere was a little chilly. For at least three-quarters of the way neither of us said anything. He sat in his corner of the tremendous car and held onto the strap and stared ahead of him. For a moment it occurred to me that it was all part of a game, and that the way this part went was that having had his suggestion about coming home with me turned down (rejection), he was now acting out the sullen "you'll be sorry" part (aloof disappointment). And all of a sudden I thought, "Doesn't he know I'm a writer—even if practically in name only? And writers are just about artists at all the *games* from the minute they set foot in a sandbox even . . ." I knew all of it . . . I had read it, written it, seen it in a hundred movies. Reject me and I'll get sullen and cold and then won't you be sorry and do anything in the world to make me the sweet, warm, considerate human being you know I can be . . . I knew it all. And yet I couldn't help wishing he'd stop being sullen-cold and making me sorry, and wishing he would be the sweet, warm, considerate human being I knew he could be.

Paula, I thought, Paula . . . Paula, *Paula* . . .

And then he said: "I'll get out of the house before you're even awake in the morning." For a second I was just glad that he had talked, and then I realized what it was he had said.

"No," I said, "you can't."

"But why not?"

The car was only two blocks away from where I lived.

"No."

We'd turned into the block now.

"Then you are afraid of me after all?"

I laughed—well, it wasn't a *laugh,* it was more like the sound of quiet hysteria.

"Whatever you want to think," I said.

The car drew up to the house. Only the chauffeur moved.

"Please," he said quietly, not moving from the far side of the car. Without answering I got out of the limousine, aware that he did not move, that he did not help me, that he barely moved his feet aside to let me pass. Aware, as the car drove off, of how little we had talked about Orin Kreedel's being fired. Aware too that—not having gotten his way—he hadn't even said goodbye.

My mother was asleep when I got upstairs. Even if she had been up, though, she wouldn't have looked for any information. Only whatever I volunteered. And I thought as I got undressed, "And you had to get to be practically an old lady before you even realized it was her way of *evening* things." She had never given me any information while I was growing up—about boys, about my period, about (Christ!) *babies*—and not *asking* any questions now.

Just before I fell asleep, I thought about Dave and his not even getting out of the car to say goodnight. I thought, "Well, all right—buster . . ." But instead of feeling angry, or annoyed, or even the-hell-with-you, I felt the same sharp stab of pain I had felt while we had been running through the park toward the restaurant . . . *while he was holding my hand* . . .

And, lying in the dark, suddenly it was all back again . . .

my father . . . Central Park . . . the biggest, the most furious one of all the whole polar bears . . . the miserable old, broken-horned yak that nobody seemed to watch except us . . . the thousand times we had walked completely around the reservoir . . . stared at the snow-frozen top of Cleopatra's Needle . . . the thousand times we had, *him holding my hand*, stalked the shadows of the hidden grapevines that stretched behind the dark, echoing bandshell in the Mall . . . Could it have been a thousand times? . . . and my father dead before I was seven?

My father—how close he seemed in the dark, quiet room now. My father—the image of all warmth, all openness, all magic.

I concentrated on sleeping, on blocking out everything—sinking, fading. But just before sleep came, there was a picture: another time . . . a year . . . a whole year after he had died . . . and a dark, cloudy street . . . and me . . . over and over again . . . walking . . . walking where we had walked. Walking alone . . . walking . . . seven . . . seven years old . . . and holding my own hand because, once you had known it, the pain you could never bear was that there was nobody—nobody—nobody when you walked in the park to hold your own lonely hand . . .

And the next thing I knew, the alarm was ringing, and when I opened my eyes it was the same little mean morning waking-up-all-alone surprise, and after five minutes of self-debate I finally conceded, under protest, that it probably was another day that had started.

ten

WHEN I FINALLY MANAGED TO MOVE MY-self fragilely into the kitchen there was a note. It said: "Dear Honey. Would you please drink your orange. It's good for you. I hope you can read this. I know I can't. Love, Vivian." I concentrated on that. It seemed simpler.

As soon as I got to work, Jane Perry told me there was going to be a meeting of the entire sales staff at nine o'clock. I said I wondered what Redford would find to say about the five editors leaving.

"He wouldn't touch it with a six-foot pole," Jane said.

"What do you mean he wouldn't touch it with a six-foot pole?"

"It's not *his* meeting," she said coolly. "Mark Post called it."

I could hear the phone ringing in my office even though it wasn't nine o'clock yet.

It was Dave Wolffe. Without any introduction, he said, "I

looked into the mirror when I got up this morning and I said to myself, "Dave, *you're* a big black panther.' "

"I can't figure out exactly what happened last night," he continued. "I have to see you this morning."

"I have to go to a meeting in a minute," I said.

"What kind of meeting?"

"Mark Post," I said. "After that I have to go over the presentation I'm doing with Steve Lunderman."

"I'll call you in a half hour," he said. "We'll have coffee."

"I probably won't be finished by then."

"I'll call you," he said, and hung up.

Whatever wild thoughts would have come into my head just then were put off, at least for the moment, by the fact that it was time to go into the meeting.

The salesmen straggled or hurried or ambled in, according to their individual styles, and—so much on the dot that it had to have been planned that way—Mark Post arrived, just as everyone else was finally in the room. He was wearing the tight little smile that had somehow vaguely started to annoy me.

He looked around the room, smiling here and there, mentioning one name and then another—"Hi, Joe . . . wouldn't mind handling the Bermuda account myself if I could keep a tan like that . . . Pete . . . how goes it; see you brought in the big one yesterday . . . good job, fella . . ." Smiling, half-bowing, "Good morning, Miss Paula." It was straight out of how to please people and influence employees, or whatever the hell the name of that thing is. Finally—the preliminaries over—he settled down into the subject of the day.

"Well," he said, "I think all of us know how small a pebble it takes to make big waves in this wild publishing world we live in. Sometimes I think if salesmen were forced to make one more call every day there'd be less time for gossip." He grinned carefully at Callahan, the red-headed liquor salesman. "It's all right, Pete, I didn't mean you—*necessarily*." It got the required and carefully counted-on laugh.

He went on. "Anyway, God knows you're going to be running into a lot of questions and conjecture in agencies in the next few days. Because if there's anybody more curious about what happens in the magazine business than magazine people it's agency people." He shifted one foot onto the chair seat.

"Actually the reason I've got you here today is to make one point—and one point only. It's damn essential that you remember that point so that when all the guff starts in the agencies you'll know exactly how to short-circuit it. The fact is," he said, practically breathing between each word, "*Today* isn't now, and never was, any one person or group of persons.

"It's always been a contradiction to me as to how an industry as vital and viable as the one we're all part of can be so fuddy-duddy when it comes to making a few changes. As for me I say changes are healthy. Sure it was a blow when Jim Canfield died . . ."

I honest-to-God thought I might be sick on the floor.

". . . but we lived, didn't we?

"Of course we lived," he said. "In many ways stronger than before.

"Well, that's it," he said, all smily teeth and charm again. "So go out. Tackle those agencies. And remember that the future of this magazine doesn't depend on any one person. Or, if it does, then it's you. So get some fun out of your job. And there's only one way to do that—and that's by putting yourself to work. There's no kick in having me or anybody else tell you to get with it. Tell yourself. Be your own boss. And for God's sake, get uncomplicated. Ours is a simple task, actually. You pick up that beautiful, fat, successful product and you talk to people about it. You either persuade them or you don't. Forget about time reports—I know old bud-dy Tom is hot about them, but forget them—just use the time smartly and advantageously. Get off your rusty-dusty and out on the street. And, after you knock over a few (customers, that is), if you don't feel the thrill of the chase, get out of this business.

"So . . ." he took his foot off the chair.

"Well, gentlemen, is anybody confused?"

The meeting was about to break up. Mark Post had already started toward the door, and the salesmen were beginning to shuffle to their feet when someone said, "Excuse me . . . I have a question."

Post turned, his smile ready, pseudo-charming, half defensive. The man who had spoken was in the back of the room; he stood up. His name was Ken McCloid; he was the Manager of the Detroit office. I guess he just happened to be in town that morning.

Short, square-built, red-faced, with a badly broken nose, he had once made All-American at Michigan. I had seen him at meetings once or twice before. While a lot of the other salesmen sat drawing triangles on yellow pads or nodding with politically wise enthusiasm—no matter what was said—he had always questioned and challenged and just generally refused to keep quiet. After one or two times you got over thinking of him as just another loud Irishman. A salesman to the teeth, he was tough, reactionary, and as honest as steel.

"I got a question I'd like to ask," he said.

Post smiled. "Go ahead, Ken. Anyway, who'd be able to stop you?"

McCloid ignored the humor, if that's what it was. "What I wanted to say is—I'm not sure I heard you right. Did you say that a magazine—that *Today*—isn't just a person . . . or, what was it you said—a group of people?"

Post said, "I'm flattered. I didn't think anybody'd listen closely enough to *quote* me." And then, "What's the matter, Ken, those automobile boys back in Detroit making you a little edgy lately?"

"That's what you said then?—that *Today*—any magazine—isn't any one person, or group of persons?"

Post smiled. "Yes, Ken, that's what I said."

"Then what the hell *is* it?"

I had never heard anyone yell at a meeting before. It occurred to me that neither had I ever seen anyone obviously angry.

Post didn't say anything for a minute. Then he sort of shook himself like a puppy that's just gotten himself pulled out of the water. "You finished, Ken?"

"No," McCloid said. "You can just bet your goddamn sweet soul I'm not finished. I haven't even started."

"Well, maybe you'd better not."

The coolness was gone. It was now almost as though there weren't anyone else in the room. Just the two of them, with all the barriers down.

McCloid said, "If you really want to stop me I guess you can. But I figure I got at least twenty pounds on you. So I guess you better decide about that now."

The quietness in the room was absolute. McCloid stood there, squat, solid as a wall. His eyes never left Post for a second.

"I started working for this magazine two years ago," he said. "I didn't know any more about magazines than I knew about canned ham or baby powder or vacuum cleaners—all of which I'd made a very comfortable living selling. As far as I was concerned, *Today* was one hundred percent the same thing—a product, right? Only easier. You had to know how to work the vacuum cleaner.

"Well, I come from Detroit—any of you know it? It's not too fancy. But then neither am I. Kimberly figured I'd get along pretty good with those automotive guys. They don't think very fancy either. They make cars and they want to sell them. Either you have something that helps sell them—or save your shoe-leather. Conversation they can get someplace else. And booze, too, for that matter.

"It worked pretty good. After all, *Today* was reaching the people who bought a lot of cars. Who the hell cared what their reading habits were. Or their drinking or smoking habits, or how many times a week they laid their wives. They bought cars—and the guys that were making me fat and rich out there sold cars. Life was rosy. You think I ever opened a copy of that magazine we all pretend to know so much about? Sure, I carried it around to their offices. Don't we all? I tossed a nice, glossy copy on their desks every month; I put them on the comp list so their wives could have a copy for their coffee table. If it was raining I carried it over my head to keep dry.

"But as for what was *in* the magazine, or *who* bothered putting it in, or any of that jazz, they could have put Mae West and Mussolini on that masthead for all I cared. I was selling advertising pages, buster—that nice, pretty, four-color stuff, separated by whatever else happens on the other pages.

"And then one day I got a call from a buddy of mine that's the advertising director at National Motors. National Motors—fourteen million dollars' worth of advertising in *Today* every year. I thought the phone would melt right in my hand.

"It seems there was a new issue of *Today* out that morning. I'd seen the cover, all right. But that was all I'd seen. Well, he'd seen more than the cover. He'd seen an announcement in the magazine of a feature that was starting next month. It was called, 'The Automobile Industry in America—Miracle—or Monster?'

"I didn't get all the specifics. The phone was too hot for me to catch the details. All I knew—because he didn't leave any doubt for me knowing—was that there was fourteen million dollars riding on the series—and I better make sure of what it said.

"I called *Today*. I'll never forget. I got Canfield's secretary. He was out of town. She told me that the editor handling the story was somebody named Orin Kreedel. Like I said, she could have said Jesus Christ and it would have made the same difference to me. Who knew about editors? Space—that's what I was talking about.

"Anyway, I asked if I could speak to this Kreedel. I remember sweating on that phone. And finally he came on. I told him who I was. All he said was, 'What did you want to talk to me about?'

"So I told him. I gave the whole smear. I told him about the phone call I had had from National Motors. I told him how tricky it would be to run a series like that. For five straight minutes I poured blood into that phone. And when I got finished he said two words: 'Drop dead.' Then he hung up.

"That series ran for the next three issues. And I guess you remember it—assuming any of you read any more than I do, that is. Anyway, Detroit remembers it. It was long, it was documented, it was honest, and it didn't pull a single goddamn punch. It dissected the automobile industry like it had never been done before. And there were no holds barred. Good, bad and indifferent, it took an X ray to the entire industry. It told what it had done to help American economy—and it told what it had done to help hurt it. It named the plusses and the minuses. It named names, handed out praise, and hit at the malpractices with a fury as cool as it was deadly.

"For one solid month I averaged about three hours' sleep a night. In a nice, plush hundred-and-fifty-thousand-dollar house, with the best mattress money can buy, I slept—more like it, I *didn't* sleep—like there was somebody outside the window with a gun . . ."

He stopped to wipe the sweat off his face. Nobody in the room budged. He was about to start again when Post said, "Ken old bud-dy—I hate to cramp your style, but in case you forgot it, the rest of us here got a few kind of important things to do"—he grinned stiffly—"so, if you've finished——"

"No." McCloid's voice would have stopped a truck. "I haven't." Then, quietly, he said, "and I haven't forgotten either, have you?"

When Post didn't say anything else, McCloid finished wiping the sweat from his face. He took his time doing it.

"I remember it was the last week of that month that advertising schedules were being set. If you're a space salesman in Detroit at that time of year you don't sleep too good anyway.

"Try to get a phone call into the Ad Director's office at National Motors? You got to be kidding. Try bribing one of the secretaries? Forget it. Sure, there were rumors about which way it was going for us. After that series there had to be. Fourteen million dollars' worth of advertising—you know how much groceries you buy in Detroit on the commissions from that? You know how much groceries you don't buy when you're all of a sudden knocked off the list?"

"Ken—" You could see Post fighting to get up spit enough just to say the one word.

This time McCloid didn't answer him at all. All he did was just look at him. Then, in a minute, he went on.

"You want to know what happened two days before those fourteen-million-dollar decisions were made? I'll tell you. Two days before I got killed, the phone rang in my office at nine-thirty in the morning. Somebody named Kreedel wanted to speak to me, my secretary said. Orin Kreedel. He was an editor at *Today*."

After a minute, quietly, McCloid said, "O.K. I'll make it sweet and simple. That same day Orin Kreedel spoke before a meeting of just about every hot-shot decision-maker in Detroit. Sure they came. They wouldn't have come for me. They might not even have come for God. But Orin Kreedel represented *Today*. And *Today* represented something they'd listen to. Maybe some of them even thought he had come to mend fences. I don't know. I don't think so. Anyway, whatever they thought, they came. All of them.

"I'll never forget that speech. I'm a pretty straight-talking guy myself. But this one. He made the steel those guys put into their cars look like paper. And what he said was simple.

" 'If you're scared of the truth,' " he said, " 'then what you better do is cream me while I'm here today. Except—that wouldn't really finish it. I guess what you really better do is find some way to kill the magazine I work for. That might not do it either. But you could try. There'd always be newspapers. And even if it didn't work—if you find that newspapers won't print what you want them to—and won't *not* print what you don't want, then it's simple, you'll just have to kill *them. And* television. *And* any two-penny paper or town crier, or you name it. It's simple. Killing something always is. It's a logical solution. Just make it your sworn mission in life to kill everybody that ever told anybody else he was doing something wrong. Including an out-of-date, unknown quantity called your own conscience. Because if any industry as tremendous as this one can't face the simple, straight truth, and learn from it, and profit from it, then I guess there isn't any answer except eliminating that truth. No matter what shape it takes.'

"I'll never forget that minute," McCloid said. "An auditorium packed with over a hundred hard-nosed hot shots—and every one of them hanging onto every word of that tough, feisty editor like it was Moses on the mountain. He was almost finished then," McCloid said. "He has a quiet voice, Orin Kreedel, plain, no frills. He didn't need any. I remember what he said was, 'And then—when that happy day comes—when you've gotten rid of those nasty little nagging voices that wouldn't keep quiet, that wouldn't let well enough alone, that refused to be kept in line because fourteen million, or a million times fourteen million dollars were involved, then, gentlemen, you'll have utopia. Only, when that day comes, that's when you better start hammering what you have to say on the walls of caves. Because it's very likely that's all the advertising space there'll be.'

"He left Detroit that same afternoon," McCloid said. "He

told me he had to get back to New York. He said his job was editing a magazine.

"What I'm saying," McCloid said, "is nobody told him to come. Not Canfield. Not Kimberly. Nobody. You know *why?* Because nobody could have made him. No more than they could have made Canfield. He came to tell them he had told the truth. But—more important than that—he came to tell them that that was what he was going to keep on doing. And that if they didn't like it, they knew what they could do."

He took a step over to Post. His voice was so quiet it was amazing you could hear. And yet you heard, all right.

"That's why I asked you to repeat what you said. And you repeated it. And now, just for the record, I'd like to repeat my question: If this magazine isn't a *person*—or a group of persons—then what the hell is it?"

After which he walked out. Squat, solid, taking his own sweet time. Detroit style.

Then the meeting broke up.

eleven

TOM REDFORD WAS STILL NOWHERE TO BE seen. I just had time to go into my office and pick up the script for the photography presentation when Steve Lunderman arrived. He was an "old buddy" of Redford's who had been given the snug job of being Kimberly's corporate head of presentations and advertising.

He had only been in the job for about a month now, and he arrived on the dot for our appointment, in which he was to go over what I had done with the photography presentation so far and make his "corporate suggestions."

"Hi, beautiful," he said as he came into the office and gave me a big hug. Jack Sheehan looked up from his typewriter and said, " 'Morning, Steve, how are you?"

"Great," Lunderman said, "or, as our old bud-dy Redford would say, 'rosy as a new-spanked baby's bottom.' "

I said, "I have the slides set up in the conference room. Do you want to see them there?"

"Sure, honey," he said, "only as old Tom would say: we better leave the door open a little to keep the landlady happy."

We started down the hallway. He was tall, dark-haired and (in spite of the semi-manic streak that seemed to be showing itself this morning) solemn looking. He was also an extreme devotee of the religion that goes If-management-is-doing-it-there-must-be-a-good-solid-reason.

As soon as we got to the conference room I began to show him the slides I had had taken, pointing out to him exactly how and where they fit into the presentation. The strangest part of it was that I found myself *selling* (in a "classy" way, but selling nevertheless) each frame of the storyboard as though my life depended on it.

As we came to the end I could feel him reaching for something to say. It was probably a bitchy reaction on my part, but it seemed as though he felt he had to find some way to justify what was probably a pretty plushy salary.

"It's very good," he said. "I just wonder . . ."

In spite of myself I felt myself stiffen. Suddenly it came home to me that, joke or not, this was the first actual presentation I had ever done, that a lot of money had been invested in it, and that, sure as anything, Redford would take his evaluation of it, whole cloth, from whatever Lunderman told him to think.

"They're no *big* changes," Steve Lunderman was saying, "but you've got to remember I've been in this promotion racket most of my life." I listened, amazed at his need to flash his credentials—even to me.

"For instance, that slide where you have the table-setting—the one where you're talking about how the *Today* audience entertains in their homes at least twice a week——"

"It's where I start to stress how they have a lot of occasions to take home movies—and to show them to people." Talk about defensiveness, it sounded as though I had a worse case than he did.

"I know," he said. "It's just—well, you know how Redford's got this campaign going to get more liquor advertising into the book."

(Magazine, you dope, I thought—books are something else.)

"Yes, I know," I said.

"Well, those glasses on the table—why don't you have them filled with Scotch or martinis or something, instead of water?"

"But they're water glasses."

"Well—wine then. You can use the same kind of glasses for both."

"But it's just the beginning of the meal. There isn't any food on the table yet."

"Doesn't make any difference," he said. "You get the photographer to take those slides over with wine."

"Sure."

It went that way for about ten minutes, after about five of which I knew there really wasn't anything to worry about. The trick in this case (outside of the presentation being at least fairly good, which I supposed it was) was just to make Lunderman feel big and experienced and wise.

Finally he beamed and said, "Well, Miss Paula" (Why the hell do people always keep calling me that, I wondered), "when I see our old buddy Tom Redford I'm going to tell him that the presentation I've seen is just about——"

But I never did hear exactly what it was "just about" because at that second there was a little knock at the door and it opened. It was my nineteen-year-old, devoted-unto-death secretary, looking wide-eyed and serious, saying, "Excuse me—excuse me, Miss Jericho—could I speak to you for a minute?"

I followed her out of the room, wondering what next, and she said, even more wide-eyed and serious than before, "Excuse me, Miss Jericho, but . . . on the phone . . . it's Mr. Wolffe." She looked at me as though I had suddenly been metamorphosed into a cross between Baby Jane Holzer and Mata Hari.

When I picked up the phone he said, "Meet me for coffee. I'm leaving my office right now."

"I can't," I said, "I'm in the middle of a meeting."

"I'll be standing right across the street by the post office. I'll wait there till you come." And he hung up.

I went back to the conference room, wondering how I was going to handle whatever it was I had to handle. Luckily Redford had arrived while I'd been on the phone, and he was obviously eager to talk to Lunderman about something, because after a few almost perfunctory-sexy, perfunctory-complimentary remarks thrown in my direction he said, "Hey, Steve, there's something I want to talk to you about . . . that tall-size Jewish Napoleon got himself some other ideas about running magazines—you want to come in now?" and they both went into Redford's office, and I grabbed my hat and coat and dashed toward the elevator.

I used the twelve-second ride in the elevator to regather whatever pieces of my cool I could manage to pick up. They weren't very many, but I stuck them together as best I could and made myself actually *stroll* across the street to the post office, where I could see him waiting.

"Hi," I said, when I'd finally sashayed across the wide avenue, self-contained, or something that passed for it, if you didn't look too closely.

"Let's walk for a while," he said, taking my hand matter-of-factly. "We can get coffee at the St. Regis."

It was strange being out at this hour of the day. It was a brisk morning, but the sun was shining, and for some reason I seemed unusually aware of all the details of the street—the on-and-off-again sound of the traffic, a gentle swish of a cab turning for a left, the rich sheen of polished wood in a tobacco shop that sold expensive pipes, the feel of my own breath coming shorter and faster than usual, so that I was aware of it rising and falling in the back of my throat . . .

When we came to Forty-sixth Street he stopped and pointed to a men's store on the second floor of the corner building. "That's where I got my first real job," he announced. "I helped open that branch for them."

And then, after we had crossed the street, pointing to a little English-type pub-place called The Rookery, "There was one whole year when I stopped there to have a drink every single night."

When we came to the corner he said, "Let's walk down this street. I used to come this way every day."

It was eerie. I couldn't put my finger on exactly what it meant to him, doing it just this way—almost like revisiting something. But why now, and why with me? "I have to go to Chicago tomorrow," he said. "Come with me."

I smiled, not knowing what else to do.

"I mean it," he said, "it'd be fun. Did I tell you about what happened in Dayton the week before last?"

"No," I said, half-surprised that there could be something left that he hadn't told me, thinking: I know more about you than practically anybody else I know—except what do I know?

"Come on," he said, "let's get you out of the cold and get some coffee." He grinned. "Don't you ever button the top of your coat, silly?" And I thought: concentrate on how he wouldn't even help you out of the car last night—how he wasn't quite so solicitous when he didn't get his way. And so I concentrated, and by the time I had, he had ordered coffee for both of us "and a croissant for you—you look as though you don't have sense enough to feed yourself."

Considering that my mission in life (I mean the one that never changes) is to lose at least three pounds, it wasn't exactly an alienating remark.

"About Dayton," he said, "I had to go to settle a paper deal we have with them out there. Brad was supposed to go with me, but he had to give a speech at a writers' conference."

I lost whatever the next few sentences were that he said. It was a syndrome I had whenever anybody mentioned Brad Rusk; I just sort of went blank with anger and it was as though in the back of my head I could hear Jim Canfield saying something, although whatever it was wasn't as clear as it had been.

Having missed whatever Dave had said, by the time I "got back" he was smiling and waving for some more coffee.

"Have another croissant," he urged.

"No—no thanks." Although I hadn't realized that in my bewilderment I'd wolfed down a whole one already.

"About Chicago," he said. "We could go up on Friday. I'll find out if there are any good plays showing there."

"No," I laughed, probably a little hysterically, "no, you won't."

"Why? Don't you like the theater? Why not?"

"Why not is because I'm not going to Chicago."

"Why not?"

"We're both repeating ourselves."

"O.K." He grinned. "I'll change the subject. We can come back to it later."

"I have to get back soon," I said. "Nobody even knows where I am."

"Who are you afraid of," he said, "Tom Redford?"

"I'm not afraid of anyone." I thought: it's pretty bad when you find yourself resorting to kindergarten statements like that one, kiddo.

Suddenly he laughed. "Did you see *The Wall Street Journal* this morning? Carried practically every word of Bob Moshier's speech to the bankers' convention in Washington yesterday. 'Profit predicted in the next quarter, says optimistic head of Kimberly Publishing Company. Bob Moshier's prediction received warmly by bankers' group.'" He laughed. "Big Bob . . . always did enjoy the limelight in his eyes."

"Don't you like being interviewed?"

"Me?" he said. "The day you see me getting any personal publicity is the day you'll know I'm on my way out."

"Brad Rusk gets quoted a lot."

"Brad's different," he said. "He's the greatest editor I've ever known."

"And probably the only one." Except that this time the remark was for my ears alone. You don't find me making two brave statements in a row.

"Did I tell you I voted both of us a raise?" he told me. "Brad and me."

"No," I said, thinking, "Well, at least you're not having any problem thinking there's anything *nice* about him this morning. I had that all neatly in place. Wolffe: villain. Me: Jane (alias Joan of Arc).

"I had my dead dog dream again," he said.

"I thought your dreams were all about being king."

"Not this one. It's about a dead dog. I saw it once with my father. One of the few times I ever saw my father, as a matter of fact." He smiled.

"Did he die when you were very young?"

"No," he said. "It's just, well my mother and aunt managed the finances of the family. They controlled the money. From the old country. My father didn't have any money or prestige. *Ergo*, he was nothing. They used to send him out of the house when any of our relatives came. After all, my mother and aunt were Sephardic . . .

"Anyway, this one day my father took me for a walk. I don't remember what we talked about. All I remember was how we walked. God, we walked everywhere. And how happy I felt. I guess I'll never know how he got me away from my mother and my aunt that one time. The poor old man—no money, no power. Who knew what terrible inclination to failure I might have picked up from him in those hours together?

"Finally we started walking back. That was when we saw the dead dog. It was in the doorway of an old closed hardware store around 110th Street, I guess. Anyway, here was this dog, kind of lumpy yellow color, lying on his back and his feet sticking up in the air. I didn't even realize the dog was dead. That was when my father said, 'I once had a philosophy, but my life ruined it.' It was that minute I knew for the first time that I didn't have a father."

After a minute I said, "I have to go back now."

"Come to Chicago with me."

"No."

He smiled. "You have to. I could fire you."

I didn't say a word. I just looked at him.

"I didn't mean that about firing you," he grinned. "You know I wouldn't do anything to hurt you. My God, I feel as though we've practically known each other forever."

I didn't answer him. We left the restaurant and started walking back toward the office. I couldn't help realizing how good I felt.

He said, "Why are you smiling?"

"I don't know."

"That's because it's just beginning."

And for just a second I shivered, hearing the words—as though someone had walked on my grave. But in a second the feeling was gone. I felt good; so I smiled.

"God," he laughed. "Imagine the same person having two impossible problems—you and Kimberly Publishing Company."

I grinned.

"I'm not going to let you go back up until you say you'll come to Chicago with me."

"I'm not coming." We had reached the building.

"You could meet Arthur Jefferson personally," he said. And I laughed.

"What's the matter, aren't you impressed by the idea of shaking hands with the Chairman of the Board of Kimberly Publishing Company?"

And then, having said that, he laughed too.

"Do you know that in Chicago at Kimberly if anybody is late more than two times a month, he has to go to Jefferson's office and give an explanation as to why? Sometimes there are at least half a dozen people waiting on a long bench outside his door."

"Like seeing the principal," I laughed.

"Exactly," he said. "As a matter of fact, there's a legend that if anyone's late more than *three* times a month Jefferson has them caned."

"I believe it," I said. "And that *secretary* of his . . ."

"You mean 'lovely Lorraine'? Yeah, she's something all right. The way I figure it, though, is that one of these days the whole ancient structure is just going to crumble apart—all the ancient methods and mores of Chicago Kimberly. For the moment, though," he said, "there's our big, stupid Aryan friend to deal with. One thing at a time."

"You mean Moshier?"

But he just dropped it at that.

"I have to go to a meeting downtown now," he said. "Promise me you'll go to Chicago."

"No, I'm not going."

We were standing right in front of the building by now. I tried to turn away, but he held me by the arm, keeping me there.

"Hello, Mr. Wolffe." It was one of the *Today* salesmen, attaché case in hand, going into the building.

"Oh—hello." But his hand didn't leave my arm.

"I'm not going to Chicago," I said. "I have to go up now." Standing there, saying the same thing over and over, I felt like a telephone company recording. Finally it got *too* Mickey Mouse.

"No, I mean it," I said, yanking my arm away. "Thank you for the coffee," and walked quickly away from him and into the building.

I was glad that at least the salesman wasn't still at the elevators.

The car, when it came, was empty, and I got into it thinking: How much gall can one person have? *Go with me to Chicago.* Except right after that I couldn't keep myself from thinking: "'*Once I had a philosophy, but my life ruined it.*' *That was the minute I knew I didn't have a father . . .*"

"Shut up," I said, "shut up!" and jabbed at the button for my floor.

twelve

AS SOON AS I GOT BACK UPSTAIRS, I WENT right to my desk, called the studio, and set up an appointment to have the slide with the water glasses changed. "In the name of my Father, I command it . . . that this water be made wine." By that time it was almost eleven-thirty, so I phoned downstairs for a sandwich and coffee and started making the copy changes in the script.

The offices started emptying around me, and for a moment I felt my usual panic at being "abandoned" that's such a standard reaction for me it's almost come to be funny. (I stress the *almost.*) Chewing the ham and cheese on rye in between sentences, I managed at least partially to divorce myself from the almost-deserted noontime offices around me, and it wasn't until I had just about finished with what I had to do on the script that I realized how alone I felt. But now there seemed to be an added dimension to the feeling. As though the recent kookiness of the involvement with Kimberly via Redford . . . via Wolffe . . . via whatever . . . was something I had become addicted to—something that made this momentary quietness of not being involved even harder to take than usual.

I thought of calling Louise. I hadn't spoken to her since the day she was home with a cold. But somehow the idea of Louise's simple, direct honesty put me off. I thought of calling my mother. But—for pretty much the same reasons—I didn't want to do that either. For a second I had a sudden swift plunge of panic, as though all the lifelines had been cut.

It was just when this feeling was progressing from strongly uncomfortable to definitely scary that Tom Redford came back. He had a little brown paper bag in his hand, and I realized that, wherever he had been, he had brought his lunch back with him. "Chicken and bacon on rye with plenty of mustard and ketchup," he announced, waving the bag at me, " 'cept, of course, it'll read more like luncheon for two at the Voisin, or one of them classy places, for the old expense account." He laughed. "I sure do like some of the fun and games things that go with big business. Why don't you come into my office," he said. "We can have us a little picnic together."

As much as I knew how crooked he was, I couldn't help liking him and thinking of him as a kind of protector. I guess it's amazing how hung up you can get on somebody who gives you a job when you've been thrown out of the last one.

"How's your little Jewish ad-mir-er?" he said, as soon as we had settled down in his office and, seated behind the big desk, he started unwrapping the production he'd brought back with him.

"Fine," I said.

"You two have yourself a nice lunch yesterday?"

"It was fine," I said. "Fine" seeming to be the word I had fixed on at the moment.

"He take you someplace classy, did he?"

"A restaurant called Monsignore."

"No kidding." He chuckled. "Getting up in the world, ain't he? You got to hand it to your little friend, he sure does learn fast."

"It was very nice," I said.

"Sure, I'm sure it was," he said. And then, so coolly he might have been talking about something that no more affected him than the weather in Alaska, "He ask you what you thought about how I been running *Today*, did he?"

"No," I said, "no, he didn't. In fact he hardly talked about the magazines at all."

And then, quietly, "What did you talk about?"

"As a matter of fact, mostly me."

"Is that right? What about exactly?"

"Oh, whether I'd been born in New York and whether I had a family here and if I was still writing and—well, just about everything."

"Want to know if you were married, did he?"

"Well, yes, he did ask that, as a matter of fact."

Redford chuckled. "See, I told you he just wanted to have lunch with you because you were a *sex-y* girl." He was silent for a moment, then with an absolutely superb imitation of coolness he said, "Didn't say anything much about the magazines, then?"

"Not at lunch. He did talk a little about them at dinner——"

"Oh, you two took dinner together too, did you?"

"Well, we didn't exactly have a chance to finish our conversation at lunch," I said, practically parroting Dave Wolffe's words. "He had to get back for a meeting."

"You two have a lot to talk about, did you?" He was still playing it as smooth as cream.

"He wanted to know about my book and what I'm writing now—and, you know—things like that."

I couldn't help myself. Whether the word was showing off, or bragging, or confiding, or even reporting, it was suddenly as though the time I had spent with Dave Wolffe had been a sort of hoarding up of facts and impressions to pass along to Tom Redford. Why? I don't know. I didn't know then. And I don't know now. The single difference being that at the time I never even stopped to ask.

"He said that Bob Moshier liked the limelight." I laughed. The effect on Redford was like cracking a whip.

"What else did he say about Bob?" he demanded.

"Nothing. He just said that Bob Moshier had made a speech and how he liked being interviewed. He said if I ever saw him getting any personal publicity I'd know he was finished at Kimberly."

"Damn lanky punk," he said. "Couldn't match shit with Bob."

With furious abruptness he scooped up the remains of his lunch and tossed it away.

"One of these days them two are going to meet head on. Then that power-happy son-of-a-bitch Wolffe is going to get cut down so fast he won't even see it coming." He moved his hand in a quick karate motion.

Suddenly, remembering something I had forgotten to tell him I said, "I asked him about Orin Kreedel." I'd be surprised if I didn't actually lick my lips, purring with still another choice goodie that I could now present to my mentor, employer, savior —God help me—friend. "Last night," I said, "I asked him about the five editors. And he said——"

"Wait a minute."

With one fast, lean motion Redford stood up from his desk. He half-raised his hand, instantly cutting off what I was saying. Then, in one quick movement, Redford suddenly squatted to examine the underside of his desk. He did the same thing with the two end tables and the three pictures on the wall.

"What are you doing?" I had an idea, but I didn't really want to admit it.

"Wait a minute," he whispered. There were still two more chairs in the room and one magazine stand. He moved quickly and quietly. Then having completed his investigation, he stood up and grinned at me a little awkwardly.

"You really think they'd bug your office?"

For a moment he seemed sheepish. "No, not your damn showoffy friend, I guess," he said. "I wouldn't put it past our esteemed editorial genius though."

"Rusk? You mean it?"

"I tell you what happened in the gentlemen's john last week?"

"No."

"There was about four of us, I guess," he said, smiling reminiscently, "all kind of taking our leisurely time, when crazy old Rusk kind of ex-plodes high-and-mighty into the sanctorium, fuming and raving and waving this long sheet of galley type and demanding that one of the editors who was at that very moment valiantly struggling to keep himself *in*-tact, right then and there lop off five hundred words from the story before he did one single other thing." He laughed so hard he hiccuped once. "I ain't hardly ever seen such a heart-rendin', lethiferous sight since I paid my ten cents on Saturday afternoons to the red-headed lady in the ticket-seller's glass cage at the Pear Street Tivoli Movie Emporium . . ."

He sat back down at the desk, and suddenly—almost as though on cue—neither of us was laughing any more.

After a minute Redford said, "No—to answer your question. I wouldn't put it past Brad Rusk to bug my office. Come to think of it I can't think of anything that I would put past Brad Rusk."

It was at that moment, for the first time considering it, that I had to admit that neither could I . . .

I stood up, suddenly needing to walk around. "It's like being in the middle of a movie. Four stars for suspense—zero for credibility, as they say in the *Daily News*."

Redford shifted his weight in the seat behind the desk. "You're wrong," he said. "Close, but no kewpie doll, like they used to call out at the fairs back in Mississippi. It's like being in the middle of an *opera*. I'd have thought any good Eye-talian broad would have seen the resemblance right off."

"Maybe you're right," I said.

"Hey—*hey* the hell with all this funeralizing. Who the hell's dead anyway? Not me, that's for damn sure. And not you, honey, God knows you're not dead, are you?" Suddenly he came around the desk and before I knew what was happening he had put both of his arms around me. "You're 'bout the most *un*dead thing I ever did see . . ."

For some reason the thing that registered with me first and most strongly was how cleverly—how *competently*—he reached behind me with one hand so that he could close the door, while still keeping the other arm around me.

I wondered what I should do. The problem was that I couldn't imagine myself doing any of the things I felt I *should* be doing, like making some funny remark that would stop him but still wouldn't hurt his feelings, or managing to "move away deftly" the way I'd read about it being done in so many books. Or even screaming. Or—for that matter—enjoying myself.

"Honey—why don't you and I just maybe take us an afternoon off the job. I mean, well, there's no real vir-tue to us *over*working ourselves." I could just barely hear the words since he had his face kind of mixed up in my hair.

He said, "I tell you what. I got a little old Avis Rent-a-Car card, Number Seven-Two-Five-Six-Oh. Wouldn't take us more than five minutes to have them drive one of them nice clean machines over here. It'd be a real nice day for a ride in the country—we could just see us some trees and scenery. Maybe even get out and walk ourselves around a little. Breathe us a little fresh air. Then afterwards we could have us a nice leisurely dinner at one of them fine wining and dining places they got out in the country there. There's this one place," he said, "one of the prettiest views you could imagine. All decorated real classy by some Oriental Jap decorator I think it was. I tell you, honey, it's a sight that'd do you good. What I mean is it's built all the way up there so you can see for miles around. Way up on a mountain. Matter of fact, that's what they called the place after—the mountain, that is—The Mo-tel on the Mountain——"

That's when the phone rang. Right on cue. At first he didn't move. Neither did I. The only difference at all was that he stopped talking.

After the second ring he said, "Damn it, where is she?" I knew that he meant Jane Perry, and that he was wondering why she wasn't picking up the phone.

When it rang the fifth time he said, "Christ, damn it." That was when he let go of me, went over to the desk, picked up the phone, and said, as sweet as sugar, "Hello . . . Tom Redford speakin' here—sorry to keep you waitin'."

I watched him as he listened for a full minute to whoever it was that had called. At first, you couldn't read a single reaction by watching his face. It wasn't until the full minute had passed that he practically radiated expansiveness; even the way he permitted himself to sit down full and relaxed behind the desk was like somebody taking a deep, satisfied sigh.

"Why, Dave, that would be just fine with me," he said. "You're sure it's convenient for you now? I mean, I know what kinda tight schedules you important executives keep yourselves to. Never a minute for your own pleasure." He winked at me, beaming.

"Fine then—I been looking forward to the pleasure of your taking dinner with me. Six o'clock be convenient for you? Fine.

Oh, by the way, Dave, I thought, if you didn't have anything else in mind, maybe you'd be my guest at a little place I started going to when I was just cutting my selling teeth. Nothin' fancy. They just kind of treat you like you were in your own home—you know what I mean, just kind of friendly and nice. It's just a little place—on Sixty-first Street—name of it's the Colony. Fine, then, just fine—I see you there 'bout six o'clock. I be lookin' forward to it. So long for now." As smooth as cream he hung up.

"Ha-*lay*-lu-ya . . . ha-lay-lu-ya, honey!"

He spun around in the chair, threw up his arms and just generally rejoiced.

"Baby, do you know what's happened—do you know what's finally happened?"

"You're going to have dinner with Dave Wolffe?" I suggested.

"Just on the surface of it," he said. "Top of the iceberg, as you might say. Baby," he said, "I'm going to get me a great big share of that there prime hog before the vultures make off with everything 'cluding the bones."

He strode toward the door, opened it and yelled out, "Jane—Jane, twitch that sassy little ass of yours and get us some glasses in here. Me and my classy promotion writer's going to have us a drink."

I said, "Thank you, but I don't think I—" but whatever the rest of it was doesn't matter, because he didn't hear any of it anyway. "Honey, do you have any idea how many times I've been trying to set up this little dinner with Mr. Wolffe? Tonight's the night I get what I'm after . . ." He chuckled and stuck both his feet up on the desk just as Jane arrived with the glasses and a pitcher of water.

"You have the liquor, I believe, Mr. Redford," she said.

"I sure do, honey," he said, "I sure do. Oh, and don't bother putting any telephone calls in for me for a while. Me and my friend's just going to sit here and booze for a while." Jane said, "Yes, Mr. Redford," and closed the door behind her. For a moment I thought I might start to panic again, but instead I found myself saying, "And what's that exactly?"

"I beg your pardon?"

"Tonight . . . You said tonight you're going to get what you're after. What *is* that?"

"I told you, baby," he said, "I told you. My share of the killing. While there's still some to get."

"But what's been killed?" I said. "I haven't seen anything dying."

"Haven't you, baby?" he asked, his voice almost quiet. "I tell you something," he said. "You may not have known that you saw it, but you saw it just the same." He poured two healthy drinks and held out one to me. "You forget I'm a country boy," he said. "I can tell when something's dead even if it ain't actually started to stink as yet."

"You mean *Today?*"

"I mean a magazine started by a particular kind of man at a particular time in the world. Well, that man's dead, honey—but there's something else, too. That *time* is dead." He took a deep swallow. "You been wondering why I didn't do anything about Kreedel, or them other editors, or any of the lunatic nonsense that madman Rusk is doing with the magazines? Well, it's 'cause I don't think it matters, honey. I think it's just a matter of time. But then, like I told you, I don't intend lasting on this job more than two years at the most. That's what my contract's for. Only believe me, in what's left of them two years I intend filling that old tomato can just as high to the spillin' brim as I can."

After awhile I said, "And tonight—that's what you intend doing?"

"You bet your sweet ass it is. What's the matter? . . . you disappointed in me, Miss Paula?"

I didn't say anything.

"Well, that's just the fact of it," he said. "Sure I'll kill them, honey. I love to see that fat slob Rusk a-roasting over a fire just as much as you would. But the fact of the matter is, it's the bounty I'm more interested in than just the pure pleasure of blood." Trying to make light of it, he said, "I'll leave that pleasure up to Bob Moshier—and he's the one that can do it."

I thought I should say something, but I couldn't think what it was. Finally he laughed; it was a little forced, a little like Tom Redford imitating Tom Redford.

"Don't get me wrong, honey," he said. "I'm still liable to pin one or two of them skins to the cabin door yet. So you won't sell me short, will you now?

"After all," he continued, expansive again, cocky as hell, "you just tell that bastard dictator that I took you out to dinner long before he did?"

He reached for the bottle again and then said, "No, I guess I better not. Got to stay bushy-tailed for my important dinner. Not that boozing ever interfered with my state of mind." He gestured toward the Jim Canfield ad that he'd had framed and put on his wall. "Feisty old bastard," he grinned. "Had to pull a damn dirty trick on us like kicking off." And for a moment I remembered again how—in spite of whatever he was or wasn't—Tom Redford's respect and love for Jim Canfield was the one complete thing that could never be cut down for real. He smiled from ear to ear, recalling, "God, it was pure poetry seeing that man cut down Brad Rusk to size . . ." He changed his mind and poured himself another drink.

"It's O.K.," he said, seeing the look on my face. "I'll have me a glass of milk in an hour or so. It's just the subject of Brad Rusk always makes me either want to vomit or get drunk." There was a look of absolute fury in his eyes as he said it.

"I thought you were only interested in this whole thing for the money," I said.

"I am. 'Cept even I can't stomach seeing a vulture chewing on something that's still got *some* life in it."

"The hell with it," he said. And then, after a moment, "Sometime when you're talking with your friend you tell him what it is he's playing with here. Four thousand people who work for this company. Four thousand people with kids to keep in diapers and pablum. You suppose it ever occurs to him to give a passin' thought to that little fact?"

Which, translated—even though I'd rather not have known—meant one man with one very real family—one scared, fuming, "fightin'-and-bitin' " but having-to-keep-quiet man—name of Tom Redford.

"Anyway I'm going to get me my fair share of the goodies from your lanky Jewish ad-mirer tonight," he said. "If that's all right with you . . ."

It was then that Redford made his last little speech of the day, the one I was to remember longest.

"You see," he said quietly, "I got no desire to be part of the *big* fight—I never did. I'm in this for two years; that's what my contract says.

"As for the *big* fight," he said, "—and something tells me it might not be too far off—sure Bob's my buddy, and he gave me this job and I'd just as soon not see him creamed. But, like I said, as far as the big fight, as far as a life-and-death feeling goes—and that's what this thing is for them others, and don't you ever fool yourself it isn't—it's just—like I said—it's just not my life style."

And then—as honest as it had come—it was over.

"Well," I said, "well, I guess I better get back to work."

He laughed. "What's the matter, honey—you just think-a some pre-vi-ous commitments?"

"I—I have to make some changes in the script," I said. "Steve Lunderman suggested . . . I have to make some . . . changes."

"O.K., honey," he said, smooth as whipped cream, "O.K., but remember—some other time—you and me, we still gonna have us our little ren-de-vous."

I could still hear him laughing as I walked out of the office, looking straight ahead, not daring to meet Jane Perry's cool and assuming expression . . .

Two minutes after I got back to my desk, Pete Larsen came in to ask me, "Eh—how did everything go this morning? I mean —well, I heard Steve Lunderman had a look at the presentation."

"Yes," I said. "I guess he liked it. He told Tom Redford it was very good."

"Oh—well, that's nice. Do you suppose you could put it on for me tomorrow morning?"

"Of course," I said. "I want you to see it."

It's all just a matter of whether you believe that bitches are made or just born that way.

"Fine," he said—my boss, that is, the man I was actually *responsible to*—and left the room.

Actually there wasn't anything for me to do. I had made the script changes. Jack wasn't there to distract by either being quiet or talking too much. Just as a side effect, he wasn't there to keep me from being lonely as hell either.

Anyway, the lunch hour had settled one thing for me. Unless I was absolutely out of my mind, even *I* had to see that this was the time to call a halt to this stupid, little tightrope dance I seemed to have found myself doing. It wasn't a bad comparison —a tightrope dance, with Tom Redford on one end and Dave Wolffe on the other. Even if you made it to one of the platforms, you couldn't be sure it wouldn't have been as deadly as falling straight down, and no net.

"Well, that settles that," I thought. "End of the game. All out. Kaput. Finis. End. That's *it*." At which point the phone rang.

Dave Wolffe said, "I have tickets for the Belasco tonight. Even the scalpers don't have any. A friend of mine knows the director. I'll send my car to your house. Eight o'clock, all right?"

I said, "No. What I mean is—I can't see you anymore. I mean——"

"I apologize for not being able to have dinner with you. I'm meeting Redford. He's been after me to have dinner with him for two weeks. I said I'd meet him, but I'm only going to stay for a drink." I could imagine him grinning. "It should be fun. I'll tell you all about it when I see you. O.K. then? The car'll take you right to the theater. Don't be late."

"But, didn't you hear me? I said I can't see you any more—I mean——"

"I miss you," he said. "I hate the idea of how much I miss you." There was a pause, and then he said, "Do you miss me too?"

And then—no pause—just before he hung up, "Of course you do."

thirteen

MY MOTHER SAID, "ENJOY YOURSELF AT the play. You look nice."

I said thanks. "Don't wait up for me."

She said, "Have a good time."

I left the apartment and waited in the hallway wondering how I would know it when it came. But there was no problem. The 7Z license plate left no doubt that this was "my car." I waited until the chauffeur had gotten out, and then I opened the door to the street just as he reached the building.

"Are you looking for Miss Jericho?" I asked.

"Yes, Ma'am."

He sort of bowed me to the car, opened the door, closed it, and then got in front. I wondered if any of the neighbors were seeing me. (It didn't make any difference to me that I hardly even knew who any of the neighbors were.)

It was funny riding downtown in this big rented limousine all by myself. It was different from being in a taxi. I tried to figure out why. Maybe because of the chauffeur's uniform. No, it wasn't that.

We drove across town on a slanty street where the kids were playing in the gutters, and one of them, a wild, Jersey-type face, maybe ten years old, blond, eyes like a tender savage, threw a stone against the car. "*Hey, rich bitch—*" I felt the stone hit the door. One of the other kids laughed like a clown. The driver's face registered fury, but I couldn't translate the curse-shape of his mouth because of the thin closed glass between us.

The limousine turned south on Park Avenue. The driver seemed more relaxed now that he was able to blend into the surroundings. I wondered what play it was; I had forgotten to look it up in the paper after I got home from work.

Suddenly I felt hungry, even though I had eaten. I wondered what would happen if I told the chauffeur to turn around and take me back home. Would it be held against him, like the nuns writing "Do Better Next Time!" across the first page of your composition?

"We're here, ma'am."

The street was packed with people. He was easing the car into a place just in front of the theater beside another limousine that was at the curb.

"I'll let you out here if that's all right, Miss." Quickly he got out of the car and opened the door for me.

"If you'd tell Mr. Wolffe I'll be here at showbreak," he said. I stepped out of the car and walked over to the front of the theater. He wasn't here yet, or, if he was, I couldn't see him. I tried concentrating on the people around me. Almost everybody was with somebody. Either that, or the person they had been waiting for arrived as soon as I noticed their being alone.

I turned to read the huge poster I had been standing in front of. The costumes were by somebody named Theoni Aldredge. I wondered what kind of a name Theoni was. It was a girl's name, I guessed. I tried to imagine what she would look like. Did all of her own clothes look like "costumes" too? Maybe she was very old and dressed like Dame Edith Sitwell. "The thing," Dame Edith said, "is that my brother and I often spend hours gazing at a single strand of grass. That is the important thing: to have time to assimilate wonder . . ."

Somebody kissed me on the cheek—and then, quickly, on the lips. "Hello, silly," he said. "I missed you."

He took my hand and held it while he made a path for us through the crowd that was stuffing the lobby. I stood against the wall and watched him as he went to the counter, checked his coat, and came back. I thought, it's like I read in the book when somebody says, "You're looking for trouble," and the man answers, "I've never had to look." But then he took my hand again and we started to walk down the aisle, and he looked at me and grinned. He said, "I've just conned a con man."

I felt the hairs on the back of my neck start to tingle. I thought of how confident and cocky Tom Redford had been that afternoon, and then I thought of what Dave Wolffe had just said, looking certain, looking sure, looking like somebody that's just won every marble in the game and then maybe more. I felt myself shiver for Tom Redford. What had happened?

Our seats were way up front. I thought: Whoever your friend is, he sure treats you right. We were almost there now. I was almost wild with curiosity. What had he meant? "I've just conned a con man."

We reached our row. One of those little, reedy women with thin red hair and papery white skin showed us which seats were ours and gave Dave two programs. He helped me settle my coat around my shoulders, gave me one of the programs, and then, before I'd had a chance at even one question, the lights went down.

I was aware of the lights coming from the stage after the curtain lifted. There was a woman sitting in a chair toward the left—an old woman with a great bony face and long white hair, and I thought, "Oh, I didn't realize *she* was in it." The scene was a living room. The woman didn't say anything or even move. And then a man came in. He was carrying a newspaper, open. He was reading it as he came into the room. There was applause as soon as he walked out onto the stage. He was the star. He pretended not to hear the applause. The old actress pretended not to hear it, either.

Instead he sort of half-shook the paper once, and then, as though reading something aloud from it, he said—I didn't know what he said. Because I *thought* he said, "I've just conned a con man."

That couldn't be right. I tried to focus on the play. I had lost the man's speech, but now the woman was saying something about the plans she had made for the man's life. It was hard to concentrate. I kept wondering whether Redford was still back at the Colony or whether they had both left together. I wondered what reason Dave Wolffe had given Redford for leaving so soon.

All of a sudden the man on the stage started to cry. I wondered what had led up to it. The woman didn't say anything. For at least a minute, nothing happened on the stage, just the man crying, softly, deeply, and the old woman sitting there as though the man had left the room, or become a ghost. Now I was conscious of the people sitting around me. One row away there was a woman in a skullcap made all of gold sequins. It was as though her head had caught fire. The man in front of me had one of those thin, reddish necks with marks that look like chicken tracks.

The woman on the stage said, "The trouble with you, Jonathan . . ."

I felt a touch on the thin gold bracelet that I always wear on my right arm. I must have had it for a hundred years. It's not much thicker than a piece of wire, and it's so battered and bent it doesn't bear more than a passing resemblance to the circle it was meant to be. I use it as a piece of jewelry, a pacifier, a decision-maker, and, sometimes, a toy. I bought it for myself a long time ago at Bonwit Teller's, and it cost me two dollars plus tax.

Dave Wolffe had taken hold of the bracelet and was moving it up and down my arm. At first he moved it very slowly, sliding it up and down my arm just for a few inches or so. I looked at him out of the corner of my eye, but his face was directed entirely toward the stage. He kept moving the bracelet up and down my arm.

There was a younger woman on the stage now. She had on a pumpkin-orange sweater and black-and-white checked slacks. She had a way of sticking out her backside at you.

The man said, "You're a vicious killer. You're a cheap, man-eating shark."

The woman laughed. She said, "Fuck you, Daddy," and stuck out her behind. The squares on the black-and-white material rounded and stretched. The bracelet moved further up and then down my arm. I thought about pulling my arm away. But it seemed silly. What was he doing? It wasn't even as much as holding my hand, really.

I was aware of every inch of my arm and the bracelet moving up and down it slowly. It was almost like being teased. But—your arm? I thought of something I had read. A funny book called *The Official Sex Manual*. "The erroneous zones are those areas of the body which are exquisitely sensitive to conginual stimulation. They are located all over the place." He was sliding the bracelet up and down—almost to my wrist—almost to my elbow—and all so slowly . . .

What had happened to Tom Redford? "There's this little restaurant we could go to—way up on a mountain . . ."

It was as though practically every muscle in my body moved with the motion of the bracelet on my arm. I tried to figure out when I had become a sex maniac. In grade school? I hadn't even particularly liked boys—except for a few wild, silent crushes. In high school? I had liked boys, all right, then, but most of the time I'd been so shy they probably all thought I was frigid or anti-social. I hadn't even kissed anybody with my mouth open until I was twenty-two. When I was in my teens and started shaving my legs I'd been positive that was the end of anybody ever wanting to go to bed with me.

The bracelet moved slowly and steadily up and down my arm. I tried to sit very still. I wasn't sure any more, even if I wanted to, that I'd be able to move my arm away.

Suddenly I had a sensation of water, of sitting for a long time in very warm water . . .

I was nine years old. I was sitting in the bathtub at home. I looked at my body, somehow divorced from it. I had never really looked at my body before. I felt far away from me, lying there in the water. It was like a mystery, one that half frightened me . . .

The man on the stage said, "The only reason I don't beat you is that I'd have to see blood." The woman laughed and showed the tip of her tongue. "Whose blood are you talking about, baby doll?" she asked.

The old woman was gone now. I wondered when she had left.

The bracelet moved steadily, very slowly, up and down my arm—up and down. Each time now, as the bracelet moved downward, it went a little farther—past my wrist—and up—and down toward my knuckles—and up—and down almost to my fingers—and—slowly—up . . .

The woman said, "Fuck you, Jewish mother-sucker," and the man lunged at her and grabbed and threw her, face down, across his knees and began to spank her furiously . . . The bracelet moved toward my fingers, encircled them, stopped—and then with one wild pull, she managed to lurch away from him, and, for a second, stumbling, she was free, but he lunged at her, dragging her back across his lean lap, and the woman screamed, "Jesus Christ God killer," but it was too late. With all the strength of his right arm, the man thrashed at her, again and again, and up and down, and again and again, and suddenly the bracelet slipped past my fingers and was off.

Instantly, before I had any idea of what was happening, the act was over. There was some applause, but it seemed awkward, self-conscious. People did not look at each other's faces. It was as though the lights had come up too soon. And the bracelet was back on my arm.

The only person who seemed cool was Dave. He said, "Shall we go outside for a while?" and even before I answered he had started to clear a path for us through the crowd.

Once we were out in the street he stopped to light a cigarette. Small clots of people were forming here and there the way they always do during intermission. I almost wanted to scream at him to tell me what it was that had happened at the Colony, but instead I forced myself to look cool and sort of bored. I was just about to explode with curiosity when he took an envelope out of his pocket and said, "Do you mind if we walk to the corner? I want to mail this to my bank before Redford sobers up. I wouldn't put it past him to try to hijack it from me."

It was a long white envelope with somebody's name and the address of The Chase Manhattan Bank. He held it casually in one hand. With the other one he took hold of my hand and we started walking down the tight side street toward the nearest corner, where I could see a mailbox.

Dave said, "The trouble with southerners like Tom Redford

is they have this mistaken belief that they can hold a lot of liquor. It's a mistake that kills them every time.

"He'd been trying to get to talk to me for more than a week," he continued. "I knew it was something pretty important to him." He grinned. "That was one of the reasons why I enjoyed putting him off." He stopped when we were halfway to the mailbox.

"In a million years you'd never imagine what that crazy man wanted to say to me . . ."

I struggled to seem nonchalant. Abruptly he said, smiling, "But maybe you aren't interested in all this petty office politics?"

For a moment I couldn't tell whether he was putting me on or not. It could be the greatest cat-and-mouse game of all time. But some instinct told me that, as crazy as it all seemed, his question was a simple, straight one.

"No, it's interesting," I said coolly.

"Well," he said, taking back my hand, starting us off again, "he had me come over to the Colony, as I told you. I think he wanted to impress me. Anyway, we started drinking. He's not as good in that department as he'd like to believe he is. I think he thought he was going to have a nice, drawn-out evening. When I told him I had to leave in an hour it took him off his stride a little. He had a proposition to make. There were only two points to it. And I wasn't to say anything until he had finished saving both. The proposition was this: I was to double his salary, starting right away. And, in exchange, he'd double the advertising revenue for *Today* in the next year. It was as simple as that." He stopped, almost as though he expected me to say something. I was thinking of Redford's "I'm going to get me a great, big share of that there prime hog before the vultures make off with everything including the bones . . ."

"By this time I could see the sweat on his forehead," Wolffe said. "It doesn't become a dignified southern gentleman like Tom Redford to sweat. They're not bred for it."

Again he was silent. I could feel a resentment in him that was more than a million years old; I couldn't even begin to grasp all of its subtleties.

"What did you say?" I hoped that my question didn't betray any of the desperation I felt to know what had happened.

"Just one thing," he said. "That I'd go along with the proposition. With one stipulation. The right to cancel his contract any time I wanted to." We were almost to the corner. I thought of Redford—"I got me a contract for two years. In that time I'm gonna get me that tomato can as full to brimming as I'm able . . ."

Dave Wolffe said, "It's official. All written out." He laughed. "I even got two of Redford's Colony waiters to sign it as witnesses." We were at the corner. I watched as he gently dropped the letter into the mailbox and then moved the flap of the box back and forth two more times just to make sure the letter was definitely in.

Then, taking one step backward, he grinned, and patted the flat side of the squat, blue mailbox. "There goes Tom Redford," he said. "Next step . . . Bob Moshier." Then he took hold of my hand again, "Come on," he said. "We don't want to miss the second act of this brilliant disaster."

When we got out of the theater the limousine was—as promised—waiting. I had even less of an idea about what had happened in the second half of the play than I did about the first. I had spent most of it making up my mind that, wherever we went for a drink or whatever after the play, I would tell Dave Wolffe I definitely wasn't going to see him again.

Still, I couldn't help thinking how nice it was having a car waiting for you to get into when you came out of the theater. It occurred to me that practically every other time I'd ever gone to see a play it had always been on a freezing-cold snowy night when the buses were either on strike or the nearest subway station was about ten blocks away. I thought how easy it could be to get used to something like this, and I'd no sooner thought it than Dave said, "I sure enjoy this more than I used to riding the IRT."

For a moment, sitting next to him in the car, I felt the backs of my legs start to tremble and my head get fuzzy, almost as though I'd been drinking, and I thought: just as soon as we get to the restaurant—or wherever we're going—the very second we get there, I'll tell him. But for good this time . . . At which point he kissed me lightly and said, "I didn't tell you. Brad Rusk and Bob Moshier are waiting at the airport for me. We're taking the company plane to Chicago. There's a meeting at eight tomorrow morning. I tried to get them to leave later." He looked straight at me. "I miss you already," he said.

"But—we have to talk."

"I'll call you as soon as I get back," he said. "I'll call you from Chicago. I'll call you every hour on the hour. I'll see you the minute I get back."

"No," I said, "I mean—that's what I wanted to talk about. I can't see you anymore."

"Don't be silly," he said, squeezing my hand and then holding it. And then, as though in a second he had completely forgotten what we were talking about, he said, "I wish tomorrow were over . . ."

I didn't ask him what he meant. I had a funny feeling about the meeting he was going to. It was as though I could sense that there was something special about it—something he was worried about. It struck me that I had never seen any sign of his being worried before. Even now he seemed cool, not really concerned. To a stranger he'd seem completely self-possessed, maybe even a little uninterested. Except that—in some crazy way, in almost no time at all—I wasn't that anymore—I wasn't a stranger.

He said, "I had a long meeting with Mark Post today. I still haven't made up my mind about him."

I wondered what he meant—Mark Post—cool, uninvolved, go-along-to-get-along Mark Post, Orin Kreedel's best "friend," great admirer and disciple of Jim Canfield. I thought, and wondered what it meant, and itched to ask, and instead sat quietly and said nothing . . .

Finally, seeing my street coming nearer, I started to say, "I mean it. I can't see you——"

And his answer, not arguing, just smiling, saying quietly, "All right. We'll talk about it next time. I promise you." Then, at my house, kissing me gently, simply, saying, "Miss me," and helping me out of the car and waiting quietly until I got into the house. And even as I started up the stairs I could hear the smooth engine beginning to move, the car starting, moving off toward the airport. Dizzy, bewildered, and, as I turned the key and came into the apartment, my mother, still not asleep, saying, "Paula, is that you? Was it a good play? Did you have a nice time?"

fourteen

I WAS AS TENSE AS A WOUND-UP SPRING when I got to work the next morning. I wished the office would fill up right away. I wished the coffee wagon would come. I finished looking at the paper quickly so I could put it in Tom Redford's office before he got there. I didn't want to have to see him this morning.

He was already seated behind his desk when I walked in.

"Oh, good morning," I said, trying to deposit the paper on his desk and get out without saying anything else.

"Thought they could buy my soul for a lousy thirty thousand dollars . . ." I had never seen him this way before. There was no laughter behind the acid this time, no half-dirty, half-funny remark putting everything in its place. All you could see was a scared, vulnerable, angry man.

I realized that he was waiting for me to say something, and for the first time it occurred to me to wonder whether he knew anything about my seeing Dave Wolffe the night before. There was no way of my even guessing.

All my compulsions pulled me to do what I had always done before—to tell Redford exactly what had happened last night, right down to the smallest detail. But for the first time something kept me quiet. I nodded, put the paper down on his desk, and left the office.

The morning was a mess. I spent it trying to digest a report that had come in from our research department. It was entitled: TODAY HOUSEHOLDS HIGH USERS OF BRAND-NAME CAT FOOD.

In the middle of trying to make some sense—or even workable nonsense—out of the whole mess, I saw Lunderman walk down the hall to Tom Redford's office. There was no slightly salacious chit-chat for me today. He had obviously been summoned by Redford, and he had obviously come running. The door to Redford's office closed behind him and stayed that way for the rest of the morning.

I spent my lunch hour walking around Bloomingdale's. I walked through their decorator rooms and pretended that I lived there. It's what I call my fantasy therapy. What I do is move into the rooms, in my head, that is, except that I redo them to suit me. And my family, of course. (After all, it's no fun fantasizing if it's only going to be about *furniture.*)

Except today I wasn't really with it. After about ten minutes of dawdling I left the store and walked back to work.

Jane Perry was just going to lunch as I reached the building. She said, "What the hell do you suppose is happening in Chicago anyway?" I stopped and said, "Why? What do you mean?"

"That's it," she said, "I don't *know.* Only that there've been all kinds of repercussions going on since about ten o'clock this morning. I don't have a single goddamn detail. All I know is that Rusk and Wolffe practically exploded a meeting there this morning. Well," she said coolly, "I better get some sustenance into me. I'll probably need it this afternoon."

I don't know what kind of chaos I expected to find when I got off the elevator, but the place was practically deserted. Tom Redford's door was open now, but the office was empty. Jack Sheehan wasn't there either. I began to think how much smarter it would have been if I'd stayed in Bloomingdale's. At least there were people.

I was passing the office of Helen Walsh, the office manager, when I heard the drone of conversation punctuated every so often by Helen's wild, sort of half-hysterical laugh. At first I didn't pay any attention, but, when I got into my office, I could still hear their conversation through the wall.

The other person was Rudolph Hecht. He was evidently so carried away with the story he was telling that he let his voice rise high enough for me to be able to hear it even without trying.

Helen said, "But do you mean to say this all happened in Chicago just today?"

"Evidently both Rusk and Wolffe must have expected it to be a great coup for them. Why else would they have had me at the meeting? Obviously they thought they were going to be completely successful, and they wanted to make sure that they'd get all the possible publicity from it they could."

Helen said, "And you mean to tell me you didn't know what any of this was about until it actually happened?"

"Not a whit," Hecht said, "whit" being one of his cutesie-type words.

"I did know that the meeting of the board had been requested by Rusk and Wolffe, but I didn't have any idea why. Obviously none of the board did either. As for Moshier, heaven knows for what reason he imagined it was called. First thing in the morning, no less. Half those old codgers on the board looked as though they were still beddy-bye. You should have seen them sort of falling into their respective places . . . some of those old geezers have to be at least eighty. God knows how many years it's been since they've read anything—much less a Kimberly magazine. Anyway, as I was saying, they all sort of fell into their slots around the table. Then they started their little game of shuffling the pads and pencils around. About five minutes, after everyone else is in place—the old guys looking sleepy, Moshier looking sort of super-salesman, but a little bewildered, as though he wasn't quite sure yet what he was supposed to sell, Rusk and Wolffe both looking sort of grim and determined—Jefferson arrives. He's always the very last one in. Sometimes I think he waits in the men's room just to make sure he's timed it right . . ."

"I must say he didn't waste much time," Hecht continued. "He just sort of plunked down in his place, looked around the table once, naturally without bothering to say good morning or anything superfluous like that, then he shot his finger out at Rusk and Wolffe. 'You have, I assume, some *important* reason for requesting we assemble here this morning.' You could almost hear the hoar gathering on each syllable. That is the word I meant to use, isn't it?" Helen laughed and said, "I wouldn't be surprised."

"At any rate," Hecht continued, "it seemed to be Wolffe that had been elected to carry the ball. He stood up and said, 'Sir . . . gentlemen—I thank you for agreeing to meet with Brad Rusk and myself here this morning on such short notice.' I remember it word for word. He said, 'As you know, both Brad and I have been expending ultimate effort on Kimberly's problems for quite some time now. And—having done so—we are forced to state reluctantly, but urgently, that there is a certain faction of Kimberly management which—though well-intentioned, I am sure—nevertheless has so badly handled the areas allotted to it that we, both Brad and I, were forced in all good conscience to bring this matter to your attention as soon as possible . . .' "

"Who were they talking about?" Helen gasped.

"Exactly my question," said Hecht. "Exactly everybody's question, I suppose. Because just at that point, Jefferson stepped into the picture. He cut right into one of Wolffe's sentences. He said, 'Am I to assume that the accusation you are about to make is directed against someone present?' For a minute nobody said anything. Rusk looked at Wolffe and Wolffe looked at Rusk and—well, finally, Wolffe, very quietly, said, 'That is correct, sir.' "

At that point Hecht stopped speaking and I heard him striking a match to light a cigarette. I was only barely breathing. After a moment I heard Helen Walsh say, "Well, tell me—who said what?" And at exactly that moment my phone rang. I was so jumpy that I almost screamed. As it was, I was so scared that I didn't pick it up until the third ring, and by that time my hand was shaking so much the receiver kept hitting my teeth.

"Hello," a woman's voice said, "may I speak with Miss Paula Jericho, please?"

"Yes," I said. "I mean, this is she."

"Miss Jericho—Miss Jericho, I'm calling from Chicago. I have a message for you from Mr. Wolffe."

"Mr. Wolffe?"

"He said to tell you he's in a meeting. He could only get out for a minute. That's why he didn't call you himself."

"Oh—I see."

"Miss Jericho, he said to tell you that he'll be back in New York before five. He said he would like to have a meeting with you, and if he should be a few minutes late getting back, would you wait in your office for him. He'll call you there——"

"Oh, but I can't—I mean—I have a meeting, too."

"I beg your pardon?"

"Mr. Redford, the publisher—Mr. Redford—he's called a meeting of the Promotion Department. For this afternoon. Five o'clock. So—well, you can see—I won't be able to see Mr. Wolffe." The phone was blank for a moment.

"I was just supposed to give you the message," she said. I wondered whose secretary she was and what conclusions she had come to about this whole thing.

"I mean—well, I'm not even sure I'll see him again," she said. "I mean—well, I might not be at my desk when he comes out."

"Well, can't you leave a message for him? I mean, can't you make sure he gets it?"

"Yes—yes, I suppose I could," she said. "That's to tell him you're going to be at a meeting—at five o'clock? Is that right?"

"Yes—yes, that's right," I said.

"Oh—well, I guess I could do that. Goodbye, Miss Jericho."

"Goodbye."

And that was the end of my communication with Chicago. And also my participation in whatever Rudolph Hecht had been saying in the next office. It was quiet now—it had evidently all been said, and I still didn't know what had happened in Chicago that morning. And just then something else hit me. The immensity of my latest stupidity. Just because I'd made up my mind that it wasn't possible for me to see Dave Wolffe again I had made up an entirely imaginary meeting. At an entirely imaginary time. And—what was worse—with an entirely real publisher. A publisher that Dave Wolffe didn't have to do anything more with than just call in order to find that I'd made up a completely phony meeting. Which was—considering everything—exactly what had to happen.

Except for one small possibility. Very small. If I got Redford to go along with the story . . . You've got to be sick, I said to myself. You absolutely have to have gone off your rocker one hundred and ten percent. You know that, don't you? . . . It was, at the best, impossible. I decided to try it.

To make things worse, when I got to his office, Redford was already back from lunch. That didn't even give me a chance to regain my sanity, realize it was an insane plan, and give up the whole idea. Because by the time any one of those things might have happened I was already standing in his doorway and he was saying, "Come in, lady. You got something on your mind, have you?" He was smiling.

I handled it the same way I dive. Gracelessly, fast, running all the way, knowing that if I stop for a second I'll be rooted on that springboard for life. The only thing is to take one deep breath, run as fast as you can, close your eyes, and jump. "Dave Wolffe called and left a message for me to see him tonight. I told them to tell him I was going to be at a meeting. Your meeting. Would it be all right if I was here when he called? I have something I want to tell you, anyway."

There's always that moment, no matter how fast you've run, when you finally smack the water. That was the way I felt right then. Run . . . fall . . . smack . . . hit—and then what?

"Sure, honey," he said, "you do that. You come in here at five o'clock and we'll have us a little drink together."

"Oh—oh, well, thank you," I said. "Thank you very much."

"Not at all. It's my pleasure." He was still smiling when I left the room.

About three-thirty Jack Sheehan came back from lunch. Drunker than I'd ever seen him. I didn't say anything. Instead I sat facing in the opposite direction. I typed out my name and address over and over again. The typewriter made a lot of noise. The last thing in the world I wanted was conversation.

"Damn professional con man," he said, slumping into his chair.

I typed faster. This time I added my phone number to the address.

"*You* trust him," he said, "don't you? You trust that conniving southern operator . . ."

I didn't answer.

"Don't you?" he said. "You do, Paula, don't you——"

I stopped typing. I still didn't say anything.

"Well, *I* don't," he said. "You want to know why I don't? Because *he* trusted me. That's what he said when he hired me. 'I know you're a boozer and you can't keep a job, but I'm going to take a chance. I'm going to *trust* you.' Well, I'll tell you something. Tom Redford's ways of trusting you can drive you to drink faster than anything else you could possibly imagine. You know why? Because he never lets you forget. Every day—every time you see him—every time you pass him in the hallway—every time you get into the same elevator with him—morning, noon or night—he never once lets you forget how much he's *trusting* you . . ."

"Jack," I said, "Jack, why don't you go home?"

"No," he said, "no, I can't."

"Why not? It's after four. Why can't you go home?"

"Because," he said. "Because Tom Redford trusts me . . ."

I said, "I guess you never *could* trust somebody who keeps saying he trusts *you*."

"Pretty cynical philosophy for a gracious young lady." Redford stood in the doorway. "Could ruin your whole personality if you lived by it. Don't you agree, Jack?"

He came into the office, obviously in his "paternal publisher" role, undraped himself into the chair next to my desk, and said, "Well, how's everything going in this here hot-shot promotion department?"

"Fine, Tom," Jack said warmly—too warmly. "How's everything going in the publishing department?"

"Oh, as well as can be expected," Redford said, smiling. He chuckled, "How about that old buddy of mine, Bob, though—ain't he the greatest?"

"You mean Bob Moshier?"

"Sure," he said. "Now there's a *winner* for you . . ."

"I didn't think you were doing too badly for a southern boy yourself, Tom," Jack said. I tried to think of something to cut into the conversation with, just to keep Jack from hanging himself completely, but I was too mixed up to think of anything to say.

"Didn't you hear about what happened in Chicago today?" Tom asked.

"No," Jack said, "did something happen in Chicago today?"

"Just another Moshier victory, that's all . . ."

"Is that right?" Jack said. "In what way?"

"Oh, it don't especially matter," Redford said. "Just a matter of separating the winners from the losers. You know who they are don't you Jack? the losers, I mean . . ."

"I guess the ones that aren't on your side, Tom. Wouldn't you say that was a pretty fair way of separating them?"

"It'd do," Tom said. "Well," he turned to face me, "you being mighty silent today, Miss Paula. Contemplatin' your future actions, are you?"

"I guess so," I said.

"Play with fire, you're apt to get burned. That's a saying my daddy used to quote to me," he said. "You heard of it, have you?"

"Seems to me I have," I said. He started to say something else, then turned back to Jack instead.

"I saw that gift-giving presentation you did," he told him. "You happy with it?"

"Not completely," Jack said. His face was white; even with his hands hidden under the desk I could imagine them shaking.

"The salesmen like it," I said. "Two of them are showing it at Young and Rubicam on Monday."

"It's amazing how little it takes to please some people," Redford said.

"Does that mean you don't like it, Tom?" Jack asked him.

Redford laughed. "Jesus Christ, *I* wouldn't know. That's Mark Post's department," he said. "He's the advertising director. If he's satisfied, I'm satisfied." He casually undraped himself from the chair, "I just wanted to know what *you* thought of it, is all . . ."

He moved slowly to the doorway. "Well, I guess I better get back to my publishing job and let you people get back to whatever it is you were doing. You take it easy now, Jack," he said, "ain't nobody ever proved a tiger can change his stripes overnight. It's the trying that counts—you remember that . . . "

Neither Jack nor I said anything right after Redford left. Probably because the atmosphere of friendly doom he had created was so strong it was as though he had never left at all.

The afternoon was wearing on. Much faster than I wanted it to. Every once in a while I thought of the whole kaleidoscope of Redford and Wolffe and Moshier and Post, and I felt sick to my stomach. It was one of those days that—no matter how you try rearranging the pieces—you just can't see the end of it.

The phone rang and it was Redford. "Get yourself in here," he said. "I want Mark Post to talk to Sheehan, and I don't want you around there."

I said, "All right," and hung up the phone, which promptly rang again.

"I'm back," Dave Wolffe said.

"You mean in New York? But I only had a call a little while ago . . . You were still in a meeting."

"No, I'd left already. Hey, I have to see you." I waited for him to tell me what all the secretive mishmash going on in Chicago today had been.

"I heard there was some kind of a big meeting in Chicago."

"Oh, that—it was just a dry run. I'll tell you all about that when I see you. Why don't I meet you at the St. Regis at five-thirty? Is that O.K.?"

"Oh, but—didn't they tell you, I have this meeting . . . I left a message."

"Oh, I got that," he said. "What does Redford know about promotion anyway? What's the meeting for?"

"I don't know," I said, tempted to say "to give me something to do so I won't have to see you." "Anyway," I said, "it's almost time now. I—I mean we—we all have to go into Redford's office."

"Don't you have any idea how much time it'll take?"

"No, no, I don't—probably long," I said. "It'll probably take a long time."

"That's O.K." he said coolly. "I have some work to catch up on here. Just give me a call when your meeting's over. I'll be at my desk."

At which point, having hung up the phone, I wondered how I had managed to make a day that had started out pretty impossibly into an absolute disaster.

As I left the room, Jack Sheehan was still sitting staring down at the drawing-board in front of him.

Walking down the hall to Redford's office, I saw Mark Post

standing in the doorway. Redford was obviously giving him some last-minute instructions; I heard him say, "You do that now, all right? He's right there in the office. Nobody be disturbing you." I saw Post nod half-smilingly, half-quizzically, as Redford said, "I'll keep our little Miss Paula here out of your hair. Fair 'nough?"

Once I was actually in Redford's office, I didn't know what to say. It was even worse remembering that I was the one who had set up the meeting—"to tell him something."

"Well," he said, "what was it you wanted to tell me?"

Stripped of my customary opening-conversational defense of at least thirty seconds of beating around the bush, I resorted to the only alternative I could think of—I plunged right in. "Well, what I wanted to say is that I want you to know that I don't intend seeing David Wolffe anymore. And I wanted you to know that my decision is based on reasons of my own—what I mean is it's not just because I don't like being put into the position of knowing via Wolffe everything that you're going to do three hours before you do it—and not just you, a lot of the other executives on the other magazines too. Anyway, that isn't the whole reason. Anyway"—sputtering off—"I wanted you to know."

For a whole minute I thought he might not say anything at all. Then, all of a sudden, he threw back his head and yelled, "*Shi-it!*" and started roaring with laughter. When he stopped he said, "Shit, honey—I couldn't care a damn who knows what about me and what I'm planning to do. They thought they had me over a barrel with my bare ass waiting for them—well, I guess what happened today takes care of that all right."

I looked at him. I don't think he really had any idea whether I knew what he was talking about or not; I didn't give him any hints one way or the other. "As for you knowing what I plan to do on *Today*. I couldn't care less about who knows what—including the mail boy. As for the other nonsense you were talking about, I think you're a damn fool if you give up the opportunity of going to all them fancy restaurants and theaters and places that Mr. Wolffe can well afford to take you to. Besides," he grinned, "why don't you stick with it—you probably end up being *Mrs.* Napoleon a year from now. The way I look at it our friend's just about ready for the *East* Side——"

"He happens to have a wife and two children——"

"Shit, honey," he said, "how come you're making it sound as though you didn't have any of that information before?"

It was a good question. "What did happen today in Chicago?" I said.

"Your friend, Dave Wolffe, and his buddy Mr. Rusk, tried to convince the board that they could run the company better than Bob Moshier. Right there, bang, bang, they shot their whole immense, imaginary bolt."

"But what happened?" I said.

"Oh, the board just sort of spanked both of them a little and told them to make up and be friends again and sent them all on their way home. That's the way Chicago Kimberly thinks, you know."

"But—they just stood up and tried to get Bob Moshier fired?"

"Something like that," Redford laughed.

"But—what's going to happen now?"

"Nothing," Redford laughed, "nothing at all. Haven't you been listening again? They'll just all come home—or they're probably home already."

"Dave Wolffe just called me from his office. They're home," I said.

"Well, it should be in-ter-esting hearing what victorious little goodies he has to tell you tonight. Beats me how he can translate this one into a plus for himself, but I've no doubt he'll give it a try."

After that we both just sat there for a minute and said nothing.

"Damn chiseling crooks . . ." For a moment his anger, hidden until now, flared into fire. "Thought they could just walk into that meeting of the Board and steal themselves the whole company. Bastard robber barons . . ." And then he laughed, elated again. "They brought all of their little tricks out into the open—and they didn't win themselves a goddamn thing—except maybe now Bob'll be onto some of the snakes he's been working with."

"You don't think he had any idea before this?"

"Why should he?" he said. "You running a whole company, you don't have a chance to go poking your head into every bin and basket every day."

"Will he fire them?" I said.

"Who? Oh, you mean will Bob fire those other two? I doubt

it," he said. "No reason to, now that they've been smacked down and put in their places." He grinned like somebody who's just killed a bear.

"Well," he said, "what about you? What are you going to do about your friend?"

"I'm supposed to call him when our meeting's finished—you know, the promotion meeting I said you were having?"

"What are you going to do?"

"Tell him I'm not seeing him again. Look, I told you, I know that maybe I didn't make myself one hundred percent clear, but I'm not going to see him again. I don't intend to."

"Don't you think you ought to at least tell him face to face?"

"You mean—see him—and tell him . . ."

"That's right," he said, "I think you owe him the courtesy of that."

Just at that moment Mark Post poked his head in the door. He was wearing his usual polite little half-smile. "Well," he said, "goodnight you two, whatever it is you're plotting." And then, as though perhaps he might have said the wrong thing, he recovered by laughing and saying, "Although I think we probably had enough plotting at Kimberly for one day as it is."

"Yeah—how about that?" Redford said, "wasn't Bob great—wasn't he just *glorious?*"

"Only a southerner could get away with using that kind of a word to describe our super-salesman friend, but I guess, yes, you might say *glorious* was the word."

"You have that little chat with Jack Sheehan?" Redford asked him.

"Yes, I did," Post said. "I told him it was a fine presentation and how valuable it would be to us."

"That's fine, just fine," Redford nodded.

"He seemed pleased," Post said. "I think it was a good idea on your part. Well, as I already said, good-night, you two schemers."

As soon as he had left I said to Redford, "You mean you sent Mark Post in there to tell Jack it was a *good* presentation right after you had practically told him how absolutely lousy it was?"

"Listen here," he said, "just because the presentation you're working on happens to be a romancy one—and, anyway, those simple guys out there think they *have* to like whatever you do, because of me—don't think that the first time you turn out a real clinker you aren't going to feel as snake-belly low as Sheehan did in there today. Remember—we can any of us fail—even top niggers sometimes."

"So you sent Mark Post in to tell him it was good?"

"I sent the *advertising director* in there to tell him it was just fine . . ." Then he put his hands over his eyes in a tired, half-furious, half-bewildered gesture I had only seen a few times before. "Damn it," he said, "why can't that fool learn to quit the booze?"

"Maybe if you'd just—" What I was about to say was that maybe if he didn't seem like such a watchdog, but just then the telephone rang and clipped off the whole topic.

Tom picked up the phone himself, and I realized it was after five and Jane had probably gone home.

"Why hello, how are *you?*" Redford smiled into the telephone. "Well, yes, matter of fact, I *did* get your little memo about our taking that insert from General Motors into the book for April . . ."

I only heard about half of what was going on; it was when he said, "Well, I guess we'd better get together and discuss the point a little further at your convenience, Dave"—that I realized who he was talking to. And by the time it had sunk in completely, he had finished the conversation and hung up the phone.

"Well," he said, grinning, "I guess maybe you better get that sweet ass of yours over to a phone and talk to your friend Mr. Wolffe pretty soon. Specially how it seems he just *might* have been calling to make sure there wasn't still some big promotional department meeting going on . . ."

He waited for me to get up and move. I said, "What am I going to tell him?"

"That's up to you, honey," he said, "except, whatever it is, I think you're going to have to at least meet him someplace and tell him face to face."

"I guess so," I said. But I still didn't move.

"Well," he said, getting up from his chair behind the desk, "while you're making up your mind I think I'll just sashay off to the old Colony for a bit. You want to get that drink I promised you this afternoon you can just call me there—if you leave Mr. Wolffe in time, of course, that is."

I didn't say anything. But finally I moved. I went into my office, passing all the empty offices on the way, sat down, and—just as Tom Redford saluted me silently good-bye as he went past my doorway—started to dial Dave Wolffe's number.

fifteen

I STOOD IN FRONT OF THE MAIN ENTRANCE to Saks Fifth Avenue. It was where I had agreed to meet Dave. There was a display by a designer named Martha Hood in the windows. Mostly she worked in bright yellows and pinks, except for the bathing suits that were designed to look like metal. In the window right next to the door there was a mannequin that appeared to be wearing a triangle of gum-wrapper foil and two flattened-down Campbell's Tomato Soup cans. There were no prices on any of the things.

Fascinated by a bright pink dress that was cut out in semi-circles, I didn't actually see him come up to me until he said, "Hi—do you like those dresses? I'll buy them for you. Only *one* of each, though . . ."

"They're not exactly my type," I said.

"Thank God."

"Well, where shall we go for dinner?"

"I told you," I said, "all I can have is a drink." I waited for him to question it, but instead he said, "well, where shall we go for a drink then? How about the St. Regis—it'll give us a chance to walk a little. O.K.?"

I said O.K. and we started to walk. I thought then—after nothing had been said for a good full two minutes—that he would bring up the subject of what had happened in Chicago that afternoon. Instead he said, "I get all my clothes from Saks now. I don't even have to go in; I just call." And it was like seeing a little boy all over again . . . cock of the walk . . . showing off.

Finally I couldn't stand it any longer. "Well, what *about* Chicago today? How did it go?"

"Oh, that," he laughed. "That was just a distracting skirmish. The real fun comes later."

It was hard to tell whether he meant it or not. He seemed completely relaxed, even amused.

"Brad made a speech to the Board," he said. "You should have heard it—it was great."

"So everything went all right, then?"

"Sure," he grinned and took hold of my hand. "I got back to you in one piece, didn't I?" After a minute he squeezed my hand and said, "I just wish you could have seen that board. The median age has got to be over eighty-five.

"Did you know that there's one old lady on the Board who has her apartment papered completely in Kimberly covers—including the ceilings."

"What does she do when there's no more room?"

"I don't know—just glues new ones on top of the old ones, I guess. She's the one who visits the printing plant once every week. Takes a complete tour of it, top to bottom. Every once in a while one of the printers or pressmen or somebody will catch her eye, and she'll go over and ask him when his birthday is. She's a nut on horoscopes. So, if this guy happens to be born in a sign that isn't 'portunate'—for real, that's her word—she fires him, right then and there. It took two solid hours one day to explain to her why we couldn't just *give* copies of all the Kimberly magazines to everybody. 'If they *really* want them, and they're *really* poor' is the way she put it.

"There is also this old man on the board who can't hear very well but won't admit it. Rusk and I had just finished making our proposal and Moshier had just started sputtering his defense when this old guy points to Brad and says, 'What's that again? I didn't hear the part you said after you stood up.' " He grinned. "Anyway, as I said, it was just a dry run. Or maybe even a red herring." He squeezed my hand. "Now promise me you'll go to your death with that secret intact, or I shall have to kill you right here and now." I grinned back (phony) and said, "But I don't know what secret you're talking about."

He smiled. "The secret that I love you very much."

I heard the words, but they didn't get through. I reacted the way people react who get very bad news and don't bat an eyelash, because if they don't react even one percent, then they've never actually heard the bad news—it never really happened, right?

So, I sat there. Smiling coolly. *Me.* The over-reactor of all time. The one who can't even stand people saying "we" because —well, just because she can't. The one now who heard the words, but who didn't dare hear the words. "I love you." And she sat there smiling coolly. She didn't even bat an eyelash.

What I did try to do, with all my might, was to remember what this meeting was all about. To tell him I couldn't see him any more. I even tried concentrating on the reality of one of the few real things that Redford had said. That this wasn't just about *me*—that it involved something with "four thousand people with kids to keep in diapers and pablum."

We had reached the St. Regis by now. The lounge was relatively crowded, but Dave managed to commandeer a tiny banquette in the corner, and after a moment, when the waiter had come, without asking me what I wanted, he ordered us two Camparis.

"At least let me have my way *once* today?" He smiled, and I said, "Fine. I'd like to see what it tastes like anyway."

While we were waiting for the drinks he faced me and said, half-grinning, "I'm not going to tell you the big secret—but I'll tell you another one."

"What big secret?"

"No," he said, smiling, "stop tempting me. You'll find out. I'd tell it to you, only I want to surprise you."

I must have looked as bewildered as I felt, because he said, whispering it as a kind of joke, "It has to do with Kimberly."

"You mean what happened in Chicago today?"

"That was the start of it." He cupped his hand jokingly over my mouth. "Now no more prying, or I won't tell you the other secret."

I was too bewildered to say anything at all, and after a moment he grinned and said, "Do you want a hint?"

I didn't answer him and he went on, "They're round, there are twenty-eight of them, and they're black. I'll give you three guesses."

"I don't know what you're talking about."

"Oh, you nut," and he made a sort of hopeless expression and said, "It's a present—for your birthday."

"My birthday? How did you know when my birthday was?"

"April twelfth—and you're four years younger than I am."

"But—how did you find out?"

He laughed, and in a second I realized that there wasn't anything about anybody connected with Kimberly that he couldn't find out all about. And for a moment I felt very strange, realizing the whole complex machinery of the organization and how—at least for a while—it had been made to serve the purpose of his finding out this little bit of information about me.

"I'll tell you," he said, his face lit up with excitement. "Anyway I'd probably get a heart attack or something waiting for you to guess. It's a string of real black pearls. They're getting them for me through Parke-Bernet."

"Black pearls?"

"They used to belong to the Queen of Denmark, and then they were put up for auction. In Switzerland. One of my lawyers happened to mention it. Cartier's has them now. They're going to call for an appointment so they can bring them to me to look at in about a week or so. I told them they definitely had to have them for me to decide about before April twelfth."

"You're kidding," I said. "Black pearls?"

"No, I'm not kidding." And for a moment the humor was gone.

"But—they must cost a *fortune*."

"Yes," he said quietly. "That's why."

That was when I saw that whatever insanity this was—

Switzerland . . . black pearls . . . the Queen of Denmark . . . Cartier's . . . Being able to do this, just because it was wild, just because it was practically impossible. All this obviously represented something completely real and bone-deep essential to him. "You didn't mean that?" he asked me. "You don't really think I'm kidding."

And all I could say was "No."

We were both quiet then, and after a while he said, "Did Redford get anything accomplished at your meeting?"

I smiled noncommittally and said, "It's hard to tell."

"Wily bastard," he said. "It kills me that I can't even explain to myself what it is I like about him. He's so damn stupid."

"Except where he's so damn smart . . ."

"What does that mean?"

"Oh, I don't know—it's just that after you get to know him, you come to see that he's only stupid when he tries to think things through. Instinctively I'd match him against anybody you can name. It's when he starts planning for a meeting, or plotting what he's going to say, that he acts like an idiot child. Also he seems to have a very powerful death-wish."

"Doesn't everybody?"

I didn't answer him, and almost lightly, he said, "What if *I* should fail at this thing I'm trying to do at Kimberly; would you be a mourner for me?" He was still smiling, but the direct, inescapable force of the question made me feel as though the temperature had just dropped twenty degrees.

"Is that what you're doing, then?" I said, "collecting mourners?"

Immediately, the full force of his anger erupted, startling me. "I said 'mourner,'" he repeated, gripping my hand, "singular—*you.*"

"Oh." I tried to make a joke of it, but I couldn't make it work. "That clown Moshier . . ." And suddenly it was as though he was gone completely from the table at which we sat. "That fool. What does he know about magazines anyway? he's just a super-grade Boy Scout salesman."

As though he had heard the question forming in my head, he said, "I learned everything I know about magazines from Brad Rusk. And that's *everything*. Rusk would die if he couldn't be an editor." And I felt myself shiver, in some indefinable way agreeing with him.

"What does Moshier know about that?" Dave said intently. "All he's interested in is selling magazine space. Yesterday I heard someone describe him as Charlie Chaplin trying to play Hamlet. Why doesn't he see right now that he can't win? It could have all been tied up today. We gave him a chance to bow out gracefully. He could have left, and it would have seemed like his own idea. Moshier's type is always jumping from job to job anyway."

Suddenly he was back with me. He grinned; he kissed me lightly and said, "Except this way it'll be even better." And for a moment, the sheer, amused determination of him showed plain.

It was a war.

I had never really thought of it as a war before. Stupidly, I had been aware only of those times and things that affected me personally. But, now, sitting there, I saw the whole huge panorama of the Kimberly scene. It was almost like something out of an old-fashioned book. It had ghosts. It had battles. It had armies. Two of them. Those who followed Bob Moshier—like Redford, like Lunderman, like all of the people who had been hired at top jobs at Kimberly by Moshier "*bud-dies*." And then there was the other army—the Rusk-Wolffe army—with people like Fritz Mott, and the limp-wristed editor Rusk had hired and some other equally bizarre types. And I thought: where do *you* stand? Which army do you belong to? And I thought: Jim Canfield—that's why you're involved at all. So obviously that meant, for me, the Rusk-Wolffe team was the enemy. Then what was I doing sitting at this table drinking with Dave Wolffe? And what about the army that I *was* a part of? Tom Redford? Bob Moshier? Oh, they were the opposing team, no doubt. But did they really represent the side that I wanted to be on? And then, head-on, I had to see for myself that there really was no side for me. That in this special kind of warfare, the line between enemy and friend wasn't as clear as it had been in the kid's games I had grown up with. The fact was that, hard as I tried, I couldn't find *my* team. I said to Dave, "Excuse me. I have to make a phone call. And then I have to go."

"Let me call first," he said. "I'll get my car here. Then I can drop you."

"But——"

He was only in the booth for a moment. As he came out I moved to enter the booth and he kissed me, child-like, full on the mouth.

"That's the toll," he said, grinning, handing me the phone.

Safe in the booth, with the door closed behind me, I dialed information and got the number for the Colony.

It wasn't until after I'd dialed and asked for Redford, if he was there, that I became aware that Dave was standing just a few feet outside of the booth. He grinned and waved, and I smiled back and turned around, wondering whether I could talk low enough for him not to hear me. A voice came on the phone. It was slurry and southern, sort of like molasses with a lot of lumps in it. It was Redford, and he had obviously had more than one drink since I had left him at the office.

"Hell-o," he said, "hello-o, who is this?"

"It's Paula," I said, "I just wanted you to know I'm with Dave Wolffe. About that drink you mentioned——"

"Oh, well—maybe it might be better another time, honey. Half of this place tonight seems to be full of folks from Kimberly. You're not angry now, are you? I don't want you to be angry."

"No," I said, almost whispering. "No—no, I'm not angry."

"I'll see you tomorrow then, honey. O.K.? Is that O.K.?"

"Yes," I said. "That's fine."

"I'll buy you a nice meal," he said. "How's that?"

"That'll be fine," I said.

"Have a good time now with Mr. Wolffe. Enjoy yourself."

"I'm going home," I said. "That's why I called you. I told him I couldn't stay any longer."

"O.K., honey, but I think you're silly not to take advantage of a chance to enjoy yourself. But—it's up to you. You do what you want——"

"So long," I said.

"So long. I see you tomorrow, baby."

"Is everything all right?" Dave said as I came out of the booth.

"Yes," I said, "fine"—wondering how much he had heard, wondering if he had heard everything and just wasn't saying anything about it.

"Can we have one more drink?" he said.

"No—I have to go right away."

"O.K., Cinderella," he said, "the car's here. I'll drop you off."

"I can take a taxi."

"Don't be silly," he said, "come on." Helping me with my coat, paying the check, holding my arm as we went out into the street and got into the sleek, black limousine that seemed to follow Dave Wolffe around like some long, black animal—all that power, but quiet, docile, waiting for him, ready whenever he needed it. Maybe—its best function of all—reminding him that it wasn't the IRT anymore.

After he had told the driver where to go he sat back against the cushions, and then suddenly he said, "Do you always worry about your mother so much?" and I didn't answer him because I couldn't tell whether he was serious or not.

It was a long time until either of us said anything, and then I couldn't be sure that he was really talking to me. He said, "They keep forgetting that Brad was a soldier before he was an editor——"

"I beg your pardon?"

He went on, but not as if he had actually heard me: "Nobody tries to protect himself against an enemy who's already made his attack—and failed. One of the best strategies is to strike your opponent when he thinks he's already won." Suddenly I was absolutely sure that—word for word—these were exactly the things that Brad Rusk had said. The things that Dave had listened to him say—and had bought like gospel.

I trembled with the absoluteness of it, and then—like the dizzy sensation of jumping off a merry-go-round—I realized that Dave had set aside the deadly, unwavering seriousness of what he had been saying. He was back with me now, laughing, taking hold of my hand, saying, "Christ, I'll bust if I don't tell you what the whole plan is . . ." And I held my breath, and he laughed and shook his head and said, "Hell, no—it'll be more exciting to surprise you." We got to the apartment, and he helped me out of the car and kissed me and said, "If you really want to know *now* I'll tell you," and bewildered, frightened for some reason I couldn't put my finger on, I said, "No—no, I'll wait." And he said, "O.K. It'll be more fun that way. Be good now, you nut." He got in, and the car drove off.

When I got into the apartment my mother said, "Louise called. She said you should call her back."

I said, "O.K.—thanks," and she looked at me queerly, I guess surprised at my silence, and said, "Did you have something to eat, or should I fix you something?"

"No," I said, "I mean—thanks, no, I'm not hungry."

"You're not so sociable either," she said, occupied with the dishes at the sink. I didn't answer her.

She said, "Aren't you going to call Louise back?"

And I said, "Oh—well, maybe it's too late."

"Nine-thirty?" She looked at me as though I were some kind of a nut. "*Be good, now, you nut.*" And I couldn't blame her. Because ever since I'd gotten to know Dave I no longer came home and gave her a Kimberly blow-by-blow description every night. All I did at all was answer direct questions. And as I said, Vivian doesn't ask very many of those. It occurred to me that I was probably avoiding Louise for the same reason. I just didn't want to talk to anybody about what was happening. Maybe because I didn't want to have to look at it that closely.

"I'll call her tomorrow," I said.

Still in my coat, I went into the bedroom and started walking back and forth. For ten minutes all I did was pace. Then I stopped to take off my coat. And then I started pacing again. Suddenly—at some point I hadn't recognized—it had become impossible for me to gloss over my relationship with Dave Wolffe as something kooky and unreal. For the first time there was something—even if it didn't have any name I could put to it—something that was as real as anything I had known. And it was the realness that terrified me.

The phone rang.

Tom Redford said, "You silly old lady. You really did mean it about going home."

I didn't answer him. I wondered why he had called.

And then the syrupy, half-stoned voice said, "How was your cocky friend? Feeling a little *chastened*, was he?" and without a second's pause I knew why he had called. Tom Redford wanted information. That had probably been true right along. Nothing had changed. Except that now *I* knew.

I didn't say anything. Who was the one who had run to him eager and bushy-tailed every morning to recount all of the precise details? Chez Vito . . . the violins . . . the Plaza . . . the limousine. Now I didn't say anything.

"Cat got your tongue?" And then—because of the long pause —"You *did* see him, didn't you?"

"Yes," I told him, bewildered to find that even the single word was difficult—more than I should say—almost like a betrayal.

"Did you go someplace?" Usual cue for the name, the location, the food, the color of the wallpaper . . .

"We had a drink."

I could feel him waiting for me to go on. Not knowing why I didn't. Not understanding. How could he, since *I* didn't understand?

"My mother's got dinner waiting," I said, suddenly grabbing it out of the air, bewildered, more amazed even than the amazement I heard in the loud, half-affectionate, half-mocking laugh, the din of the Colony behind him, and then again the silence and him waiting for me to break it.

"Well, better not keep Mama waiting," he said finally, and I said "No," and he chuckled and said, "Get yourself some beauty sleep now. Maybe you feel more talkative in the morning," and he hung up.

After a few minutes I stood up and went back into the kitchen. My mother said, "You want a cup of tea at least?" and I said, "No. I think I'll take a bath and go to bed early."

She turned around at the kitchen sink and looked at me. As I closed the door to the bathroom I heard her say, "So go ahead, be Greta Garbo. See if I care."

sixteen

THE NEXT MORNING A MEMO WAS CIRCULATED to the entire staff. It said, "Under no conditions will any Kimberly employee make a statement to any member of the press." It was signed by Bob Moshier. Nobody had any idea of exactly what it meant, but it created instant chaos. There was one theory that it had to do with what had happened at the board meeting in Chicago the day before. There was another theory that the whole organization had been taken over by CBS. There was even a theory that the police had raided the corporate suite that Kimberly kept at the Waldorf Hotel and that the whole story, complete with naked pictures, would be in the next issue of *The Daily News*. Secretaries buzzed to each other across their desks. Executives looked enigmatic but somehow vague. One of the decorating editors kept going around, wringing his hands and saying, "But I *have* to talk to the press. Peter's spent this *entire* month arranging for me to be on Hugh's show . . .

Oh, what does it *mean?*" He said this over and over again to whoever would listen, each time wringing his hands and giggling hysterically.

At ten o'clock my phone rang. Dave Wolffe said, "It's like being in the middle of something by Gilbert and Sullivan. How are you, darling?" For a moment the word stopped me dead, and then I said, "I'm fine. Only—well, what's happening?"

"Well, to be precise," he said easily, almost jokingly, "there are two S.E.C. executives in my outer office, I have three calls waiting for me on my other extensions, and I think my secretary is either about to jump out the window or have fits. But what I really called about was to ask how you are. You seemed sort of strange last night."

"I'm fine," I said.

"The only bad part of it is that I won't be able to see *Hamlet* with you. I'll probably be here all night with this insane thing."

"*What* insane thing? What about *Hamlet*—you never said anything about *Hamlet* . . ."

"Do you know how I had to wangle to get those tickets? It's the only time Gielgud's ever done *Hamlet* on the American stage. I'll send the tickets over to your office. You can have somebody else take you." He laughed. "Whoever he is, I hate him already."

"*What* insane thing?" I practically screamed. "Would you please tell me what we're both talking about?"

"Oh—that," he said, and even over the phone I could see him grinning, amazed at the machinations of big business, involved with something now called The Magazine Game that couldn't, after all, be very different from all the games he had played before. "Oh—that," he said again. "It's just oil. It seems we own this timberland in Alaska. It only supplies us with about one third of the paper we use. Anyway, it seems we've struck oil up there. At least it's on some property directly adjacent to ours. They'll have to find out whether it extends onto Kimberly property or not. Isn't it wild?" He sounded like some kid who'd won the game, except that it hadn't been by planning or skill, just luck, just some crazy kind of no-sense luck. "News only came in about an hour ago. The stock market's going crazy. I'll probably have to fly up there tonight. I had planned on our going to the Sign of the Dove after the theater. Hey, don't you think it would be better if you asked a *girl* friend to go with you?" I could hear what sounded like mass confusion behind him, and he laughed and said lightly, "I think my secretary's about to give up the ghost. I guess maybe I better talk to those guys . . ."

"Yes," I said, "I think that might be a good idea."

"Wait," he said, "hey, don't hang up so fast. I've just had a great idea. Will you fly up to Chicago with me this weekend? They're trying out a new play. It sounds great. We could have dinner at The Embassy. They have fantastic lobster. O.K.? We can use the company plane. O.K.?"

"No," I said. "No, I can't go."

"Yes you can. I'll find a way. No kidding, it would be great. Do you know Chicago? A lot of it is very nice. O.K.?"

"No," I said. "No. I can't go."

"Listen, we'll talk about it," he said. "I'll make the arrangements. We'll talk about it right after I get this *tsimmis* under control . . ." I could hear a chaos of voices behind him in his office.

"I'll call you tomorrow," he said. "Maybe today if I can——"

"I can't go to Chicago with you," I said.

"It's a great flight. I love you, baby," he said, and hung up.

Later that morning news of the oil strike began to circulate, but even then there were fewer details than Dave had told me. The rumors varied about it being oil or some kind of ore, and the whole thing was just generally vague and hysterical.

I guess it was typical of Tom Redford to choose this time of general confusion to create some of his own. Jack Sheehan had called in sick, so there was nobody except me in the office when Redford came in at about eleven and plunked himself down in the chair next to my desk.

"Well," he said, "and what *else* is new?" a phrase of his that always drove me up the wall. I didn't answer him, and both of us were completely quiet for the next minute.

Then, in typically unpredictable fashion, he said, "I guess you think you're a real big deal just because you hoodwinked them simple salesmen into thinking you're some kind of a hot-shot writer . . ."

The comment had the expected shock value, at least for a second. Then I decided I'd try to handle the whole thing as a joke.

"Sure," I said, "me and Joyce and Hemingway." But Redford chose not to see the humor of it.

"Wait till my buddy from *Newsweek* magazine takes over. Clip West— He be moving in any day now. Soon as he gets all the odds and ends tied up at *Newsweek*. Believe me, we'll see some changes now with him as promotion director . . ."

"But—I thought we had a promotion director."

"Pete Larsen?" he said. "Oh, there'll be plenty of work for Pete to do. He and Clip'll just kind of—work *together*."

I was quiet, letting that little piece of information sink in, when Redford, almost as though I had challenged him, said, "Well, we'll just have to see—if Clip likes your work, fine. If not, I guess you'll just have to find work someplace else."

It was too much. I said, "What the hell do you think you're doing anyway? You—the one who's always preaching about 'extra effort,' and all you do is go around creating an atmosphere of fear. You expect 'extra effort' to grow out of that?"

"What do you mean?" he demanded, the voice of completely outraged Mississippi.

"You know what I mean. You keep preaching to the salesmen about 'initiative' and 'extra effort,' and what you really try to do is run the place on fear. Why do you think the salesmen go out of their way to avoid you? You just think about it for a minute. Sure, they think you're a great guy—they think you're funny—but when it comes to having a life-or-death decision about firing one of them or not, they go out of their way to avoid you. They know you're a kook—they know that any minute you might decide to do some kooky thing. So go ahead, run the place on fear—you might get an adequate performance out of most people, but adequate's the best you're going to get. As for 'initiative' or 'extra effort,' forget it."

"Well," he said, "well, we'll just see about that. If that's the way you feel—if you can't stand the *atmosphere* of *fear* I cause around here, then I guess you're just going to have to look for work someplace else." And he stalked out of the room, still sputtering.

I was in such a rage I couldn't make up my mind whether to quit, cry, scream or what. The more I thought about what he had said, the more furious I got. It wasn't just his pushing Pete Larsen out of the picture—although it was pretty clear that was exactly what he was doing. What I was really reacting to was his attitude about *me.* There was just something about the way he kept saying, "you'll just have to look for work someplace else," that riled me. I guess my once having been dirt-poor helped. Completely thrown, I dialed my panic number. But I couldn't really concentrate on what it was saying. It was something about "Teach me the virtue of mercy, O Lord . . ." but I was too riled up to really pay any attention. I saw Tom Redford pass my office, and I hung up even before the record had finished.

I tried concentrating on what I was supposed to be writing, but I kept mumbling to myself and seeing red, and finally I decided I could just as well go out to lunch early as stay here accomplishing nothing.

I went to the ladies' room and washed up, and I was just coming back to get my coat as Redford stepped out of his office. He called and started down the hallway toward me. I wondered what extra-nasty thing he had to say that he had forgotten to get in before. I braced myself for whatever it might be, and at exactly that moment he reached me, put his arm around my shoulder, and said, "Goddamn it, I never did know how to get along with *creative* people . . ."

The absolutely pure, abject humility of the statement practically knocked me off my feet. His arm still around my shoulder, he walked me back into my office.

"You're going to work out just fine with Clip when he gets here. It's just my way. Like I said, I never did know how to get along with creative people."

"That doesn't mean you couldn't learn if you wanted to," I said grudgingly. I was fighting the compulsion to forget everything that had been said before. It was hard to keep holding onto that knowledge with his arm around my shoulder and the pure southern sweetness pouring out of him.

At that point, putting both arms completely around me, Tom Redford said quietly, "No—I'm too old to change—and I hope you are too."

I guess I smiled in spite of myself. He seemed relieved.

"Going out to lunch?" he asked me, with that easy, throw-away style that always meant he was hiding his interest.

"What?—oh, yes—yes, I am."

I started to move toward the door.

"Meetin' somebody, are you?"

"No." And then, without even thinking about it, I realized exactly what had caused his about-face attitude—the sweetness—the "I hope you're too old to change . . ." It was really funny. I was tempted to tell him that I knew exactly what was going through his mind. He had seen me on the phone before. The fact was that he had no way of knowing *who* I had been calling, or exactly who I might be on my way to meet now—he had no way of knowing it might not be Dave.

I didn't really grasp how much of a new thing this was for me: learning to look at Tom Redford objectively—figuring his ulterior motives. Even accepting that he might have any.

The little counter-type place around the corner where I went for lunch was crowded with Kimberly people. And the talk was all about Alaska and the oil strike and what it would mean for the company. There were all sorts of wild conjectures, including one that the whole thing was a hoax just to get the banks to loan Kimberly more money. One man said, "I just tried to put some of my savings into it, and my broker tells me that Wall Street's stopped all activity on Kimberly for the day."

It was a strange feeling sitting in the middle of all this and realizing that I had known about it practically as soon as it had happened. A strange feeling—and a very lonely one. It was exactly that moment in which I suddenly realized with both surprise and sadness that I was—as instructed—missing Dave Wolffe.

Then I remembered Jim Canfield and how beautifully black and white everything had been for me then; and, slowly chewing my ham and cheese on rye, I said to him, "A real cop-out, right?" But there wasn't anybody to answer me and, as for me, I didn't have any real answers either.

seventeen

THE TICKETS ARRIVED AT MY DESK BY messenger that afternoon. There was a sheet of memo paper with them. On it was written: "Enjoy the play. Miss me."

On impulse I decided to call up Louise.

"Oh, hi, Paula," she said, "I figured you'd either eloped or joined the Carmelites."

I tried to keep the defensiveness out of my voice as I said, "Oh, you know—Kimberly . . . "

"Busy, huh?"

"I have a ticket for *Hamlet* for tonight," I said. "Only one, though—what I mean is I don't have one for Bill, just for me and —well, I wondered whether you'd want to go."

"Hey, that's swell," she said. "Bill's out of town this week. There's a teachers' conference in Ann Arbor. Why don't we meet early and have dinner first? It'll give us a chance to catch up."

For a second I regretted having called her at all; then I thought: what the hell's going on? This is Louise. *Louise.*

I said, "That's a great idea. Why don't we meet right after work?"

"I'm on a diet," she said, "practically all I can eat is raw meat and yogurt."

"Why don't we meet at Downey's?" I said. "It's right near the theater, so we won't have to rush. And they have great steak tartare." Not realizing until I had said it that the only person I had ever been to Downey's with had been Steve and—the hell with that. It's a big city, I thought. Are you going to keep building shrines? Grow up.

"It's at Forty-fourth and Eighth Avenue. Meet you there at about a quarter to six, O.K.?"

And she said, "Great," and then, "Hey, how'd you happen to get the tickets anyway? I thought it was practically impossible."

"Oh—well, it's a long story," I said.

All through that day—remarkably, almost strangely—nothing, nothing *Kimberly* really happened. I got to Downey's just about a quarter of six—realizing as I walked the last half-block that I hadn't been near here since Steve. By the time I was breathing normally again, I saw that Louise was already there waiting for me.

She said, "Hi, child!" (she's exactly one year older than I am), and I realized how good it was to see her.

As soon as we had ordered a drink (Louise is the only person I know who makes no bones about the fact that her favorite drink is gin—straight) she said, "Bill called from Ann Arbor this afternoon. He sends his love."

"How long has he been out of town?"

"Six days. Since Saturday."

"You must miss him."

"You better know it." The drinks came and she took a long sip and gave a deep sigh.

"Well," she said, putting down the glass and licking her lips, "*give* . . ."

I must have looked blank because she said, "Kimberly—the sordid details. I'm at least three weeks behind."

I took a drink and said, "Oh, well, you know—it just keeps going on . . ." She looked at me peculiarly, which wasn't too hard to understand, since previously I had never been able to wait to give her every throbbing detail, usually the same day it happened.

Instead of pushing it right then she said, "Well, then, let me tell you a little about the mysteries and mores of the New York Telephone Company." At which she went into a description of the latest research experiment she was involved in.

I keep forgetting how smart Louise is. She had no sooner begun to dip into, "The ratio of total communication potential"—when she put down her drink and said, "Either you've lost your job, you've seen Steve, or something's happened at work. And in any case you don't want to talk about it. So which one is it?"

In a way I felt relieved. It was as though, as much as I had resisted it, I had been waiting to talk to somebody about Dave. Somebody who mattered.

Even so, I might not have known exactly where to begin, but Louise asked, "Have you been fired?"

And I said, "No," and she said, "Then what's his name?"

"His name is Dave Wolffe. He's head of the magazine division."

"I know," she said. "You told me that much before. So what happens after you find out his name?"

I had never felt such a sense of release as I did telling Louise about Dave. There wasn't a single place we had gone or thing we had done that I didn't tell her about. It was different from telling Redford. That was sort of for a purpose, even though I didn't understand exactly what purpose. This was telling somebody I knew. A friend. Louise.

I don't know how long my recitation must have taken. When I had said the last word, the waiter came to take our dinner orders.

Louise ordered the steak tartare without even looking at the menu. I came up with the first thing that had caught my eye, London broil. As an afterthought and on impulse, just as the waiter was leaving, I told him to bring us half a bottle of red wine.

"What are we celebrating?" Louise said lightly.

"Oh—I don't know," I laughed. "Maybe just seeing each other again. It's been a long time."

She picked up the gin glass, took a swallow, and then said very quietly, "Are you going to marry him, Paula?"

"Who do you mean?" I said, like some stupe, knowing all the while, but not knowing what to answer or even to think.

"Don't you know that when you talk about him it's perfectly clear that you're in love with him?"

I didn't answer her and she said, "If you could just hear yourself when you say things like: His hair's starting to go a little grey . . ." She touched my hand. "Oh, baby"—and for a moment she sounded a hundred years older than me, a million years wiser—and kinder—and not wanting me to be hurt . . .

"He's married," I mumbled, and she didn't hear me. She said, "I'm sorry?" and I said it again, my voice like stone, perfectly even and matter-of-fact. And dead.

For a second you could see the fury strike in her eyes. And her voice had a terrible quietness in it when she said, "What did he think? Did he assume you went with the job?" She was quiet for a minute and then she said, "I'm sorry. I had no right to say that."

"No," I said, "no. It's all right," stumbling, confused. Because out of it all, out of everything she had said in just those few words, the strangest part of it was a feeling that exactly those same words had been said before. And then I remembered. I was the one who had said them. And after that there was very little for us to say, and we finished our dinner with some sort of awkward, light talk or other, and then it was time to go to the play.

I guess I've seen about half a dozen versions of *Hamlet*. The one I know I didn't see was the one I went to that night. I would hear the beginning of a speech: "To be or not to be," and then I would be back in Downey's, and it would be Louise—"Are you going to marry him?" . . . "If you could just hear yourself."

It wasn't possible, I thought. Hadn't I known way back in the beginning exactly *why* I was involved in this thing? Hadn't I hated the whole idea? Magazines . . . the big cause? the Joan of Arc thing? Wasn't that why I'd even bothered ever meeting Dave Wolffe? "There's rosemary, that's for remembrance; pray love, remember"—and then I didn't hear any more, only Louise saying, "If you could just hear yourself."

There was one intermission, and we inched our way into the lobby and then into the street. Right away Louise started talking about the play so I wouldn't have to say anything. She must have forced herself to keep on talking. After a while the lights started blinking and we moved slowly back inside.

The theater had no sooner darkened when it started all over again. "Don't you know you're in love with him? *Don't you know?*"

I thought of what he had said, "If I get killed in this place, will you be my mourner?" And I had laughed. Had it been obvious then? Hadn't I listened to everything he had said with a strong degree of detachment, amusement—skepticism, even? So when had it happened? At what point had I stopped being detached, amused—skeptical, even?

And all the silly things he had said—"You don't even know how to take care of yourself . . . You're not afraid of me, are you? . . . I miss you already . . . Don't you think you should take a *girl* friend to the play? . . . Say that you miss *me* . . ."

And suddenly I was very frightened.

Because I wasn't fighting a cause anymore—because I hadn't ever really fought it. Because I didn't want to see him to convince him of my ideas about Kimberly—but just because I wanted to see him. Because it had happened. *I missed him.*

It was just then that the curtain went up on the second act of the play. I cried as quietly as I could.

eighteen

IT WAS THE NEXT DAY THAT CLIP WEST arrived on the scene. Redford had come in early that morning to supervise the handymen who were to move Pete Larsen's desk into the same room with Jack Sheehan and me. He had kept them juggling the three desks around, trying to make it look "perfec'ly natural." He had even gone into a speech for the benefit of me, I suppose, about how it would be "more logical" to have Pete in here with Jack and me "so as to supervise things . . ."

When Pete finally arrived, he walked down the hallway to where his office had been. Only then did I realize that evidently Redford hadn't prepared Pete for the change.

I heard him start down the hallway toward where I was sitting and then hesitating, stopping, walking instead toward Redford's office. I could imagine the bewildered look of surprise on his face—of not understanding, of wondering what had happened, whether it was a joke or what.

I don't know what he said to Redford, probably something half-rabbity, smiling, and I stood up, grabbing my purse, wanting to get to the ladies' room or ride the elevators or anything, just so I wouldn't be there when Pete actually entered the room. But I was too late. Larsen and Redford were already coming down the hallway, Redford's arm around Pete's shoulder, a long string of treacle flowing from Redford . . . "handier for you in here . . . Clip'll be asking a lot of questions . . . he'll need to be right next door to Mark and me . . . a better arrangement in the long run . . ."

Redford didn't actually come into the room. He just escorted Pete to the door and then left him, by which time I was sitting at my typewriter again, pretending to be concentrating on something I was writing, acutely aware of every nuance of his reactions.

First he looked around the room like a puppy dog set down in unfamiliar surroundings. Then, slowly, he walked around the desk, stopping finally behind it. Then, tentatively, as though invading privacy, he slowly opened the top drawer. I heard his small gasp of surprise at seeing that the things in the desk were actually his—that the desk was actually his own desk—that he had, in fact, been moved out of his office, dispossessed.

I started to bang the keys on my typewriter, but no matter how much noise I tried to make the room seemed absolutely quiet. And then I heard a sound I never want to hear again in my life. Pete was sitting at the desk, his head buried in his hands, his shoulders shaking as though he were an old, old man. That was when I grabbed up my purse and went quickly out of the room, closing the door behind me. Maybe, I thought, maybe if I lived a very long time he'd be able to forgive me for having seen what I did.

I took a long time in the ladies' room, even though the lighting in there makes you look as though you were half a pound of chopped meat that's started to go bad under cellophane in the supermarket. Then I went to talk to Jane Perry. I asked her if there had been any new developments about Alaska and the oil and everything, and she said, "I keep hearing so many different rumors my head is spinning. As far as I can tell Dave Wolffe's the only one that's actually gone up there. Wouldn't you think

Bob Moshier'd be interested enough to go himself? Except considering everything that's been happening—I mean Rusk and Wolffe trying to push Moshier out at the board meeting—nothing surprises me anymore."

"Clip West is arriving today," I said. "They already moved all of Pete Larsen's things out."

"Jesus, this place would have made a great mortuary," she said. "They keep shoveling the bodies around before they've even stopped jerking."

I went back to my office finally, because I had to, and Pete Larsen was gone. I asked Jack Sheehan, "Where's Pete?" and Jack said, "I don't know. He mumbled something about having to go to a studio. He said he'd be back in time for our lunch, though."

"Oh, are we having lunch?"

"Yes," he said. "It seems as though Mr. Clip West, whoever he is, is taking you and Pete and me to lunch. Anyway, Pete said he'd be back in time. What are they doing?" Jack asked. "Are they firing Pete?"

And I said, "No—nothing that healthy." After that I didn't want to talk about it anymore. I started typing. Every once in a while I caught myself almost putting out my hand to touch the phone, except that it hadn't rung. I told myself that it was just nerves—who wouldn't go jumpy in this crazy place—but I knew who it was I was waiting to hear from.

Later that morning Clip West put in his first appearance. He made his way to his office, that had been Pete's, with all the directness of a bird dog. And in a minute Tom Redford was in there with him, the two of them laughing and chuckling with all the sticky joviality of two old American Legion members who haven't seen each other since the last convention in Atlantic City.

About five minutes later Tom appeared in the doorway with West. The first thing Tom asked, looking at the empty desk, was, "Where's Pete?"

"I think he had to go someplace," I said.

"Well, then we'll just have to start with you, Miss Paula. Clip, I want you to meet a classy promotion-writing lady, Miss Paula Jericho."

"How do you do?" I said, and he said, "How do you do," and

then Redford said, "And this here's the quiet one of the Promotion Department, the art director, also be working for you, Jack Sheehan."

And Jack said, "How do you do?" and Clip West said, "How do you do?" and Tom said, "Jack's recently finished designing a very enlightenin' presentation on gift-giving. I think you might be interested in seeing it when you get a chance."

And Clip said, "I surely would. I want to get to see *everything* just as soon as I can."

"I knew you'd want to meet with Mr. West right away, Miss Paula," Redford said. "He's the man you'll be working for . . ."

And I thought, "You dumb cluck, you can't ever leave well enough alone, can you?"

West smiled and said, "The way I see it, we all be working *together*."

He was built very tall but square, and there was something of the ex-athlete about him—the shuffling way he moved, maybe, or the way he stood—and he had a kind of soft, harmless-looking face, a sort of squashed nose and big ears, and eyes as soft as a cow's with long lashes that it seemed a shame to have wasted on a man, and from the first moment I met him I knew I wouldn't ever trust him one cent's worth.

He said, "I'm lookin' forward to our all getting to know each other at lunch today. Maybe you'd be kind enough to suggest someplace we could go, seeing as how I'm a complete stranger around here myself."

"Monsignore is nice."

Tom laughed. "I tell you this one here sure has got some fancy tastes. Apt to pick them up in sort of *peculiar* ways some of the time though."

And West said, "I thank you for your suggestion. I'm just a hayseed when it comes to knowing where to go. You two gonna have to smarten me up," and he looked, wide-eyed, ingenuous, as harmless as a butterfly, smiling at Jack and me. "So I won't be a disgrace to this fine, sophisticated magazine."

Tom chuckled and said, "I swear, Clip, if I didn't know you better you'd have me believing all of that."

West just smiled and said nothing. It was a habit of his I was to notice fast and remember long. Ask him anything he didn't want to answer, or make a joke or statement that put him someplace in the middle, and he'd let it flow right off him, just as though you'd never even opened your mouth.

Now Redford said, "Well, I guess we better keep moving, Clip. I want to introduce you around to all the guys just as soon as possible. They going to be delighted to know the kind of talent they got going for them . . ." And West said to Jack and me, "Well, I guess I say goodbye for now. I'm looking forward to our takin' lunch together in a while." And they both left, Redford grinning like somebody who has just won the sweepstakes, and West smiling, pleasant, innocuous, everybody's simple old friend.

I had to go down the hallway, and I heard Tom talking to one of the salesmen, "What Clip'll do is to serve as head of a complementing division of Promotion . . . Clip and Pete'll be on a kind of a *lateral* basis, so to speak."

Pete Larsen came back to the office just before lunch, and Tom was waiting to introduce him to Clip West.

Clip said, "It's sure nice to meet you, Pete. Tom's been telling me so much about you . . ." Pete just blinked and smiled his weak little smile and said, "Well—that's a very nice thing to say, Clip."

"We be going to lunch about twelve-thirty, if that's all right with the rest of you. I had Tom's little secretary make us the reservation. That's not an inconvenient time for you, is it? I mean, well, you just tell me if it is."

"No, not at all," Pete said.

"Well, I had Miss Jane make us a reservation at that place you suggested." He looked at me, and I half-expected him to say, "Miss Paula," but he didn't say anything.

The four of us made a funny procession at lunch time, walking over to the restaurant. Jack and I paired off and left Pete and Clip to walk over together. Most of the way nobody said very much. At one point, waiting for a light, I heard Clip say, "Ain't that Tom Redford a hot one, though? You should have seen some of the wild ones we pulled off when we were buddies together over at *Newsweek*. I tell you—he's a sketch."

At the restaurant Clip made a big thing of giving his name to the headwaiter and of adding that he was from *Today*. There was very little said among us until the waiter came and asked if we cared for a drink. Jack and Clip both ordered martinis and Pete had a Tom Collins. Almost without thinking about it, I found myself ordering a Campari. I also found myself looking around the restaurant to see if anyone I knew might be there.

After the drinks arrived things seemed to open up a little. Pete made a little half-salute with his glass toward Clip and said, "Well—welcome to Kimberly," and we all of us took our first sip to that. Clip said, "I think if I had to choose my most favorite place in the world to be working, it would be at *Today*."

"It's a fine magazine," Pete said. "Too bad you never got a chance to know Jim Canfield. He was a really great man."

Nobody said anything, and I thought that already it seemed a little out of character for Clip not to jump in with some greatly enthusiastic statement, which appeared to be the role he had chosen for himself—at least so far.

"Yes—we sure had some great years," Pete said, his eyes filling up a little, but whether from sentiment or the Tom Collins or both it was hard to tell. "I remember once we did a promotion folder—it was about the affluent *Today* family—and somewhere near the end the writer referred to these people as being 'the most unique.' Well, Jim Canfield was on that phone two minutes after he saw it. I can still remember him chewing me out—'How can we expect to have any literate readers if we don't know the bare basics of the language ourselves?' He was a good friend, though—if he liked a certain promotion piece, he'd always call and tell you. We did some real nice things."

"You sure did," Clip said.

"I remember when I just arrived at Kimberly," Pete said, "it was only two years old. Those were the good old days. We really did do some nice promotion pieces." And I wanted to kick him under the table, for the sweet and horrifying way he was defending himself—when there hadn't even been a decent open attack.

"You know that lucite piece that has the miniature of the 'space' issue of *Today* in it? We spent a lot of effort on that. Getting the shape just right and making sure the lucite was perfectly clear. Why even now—and it must be at least ten years since we did it—you can still walk into some advertising executive's office and see one of those pieces on his desk. Like I was telling Tom—we ought to do something along those lines. As it is we don't have any real quality piece at the moment that represents *Today*."

"That's a very interesting idea," Clip said. "Of course for a while I think what Monk has in mind is using most of promotion's budget and time to build up the identities of the editors. Did you know they were doing a profile on him for the *Times* next month—all about his ideas on running the magazine and how he's planning on making some changes, you know, get rid of some of the fuddy-duddy notions people been getting about the book . . .

"Monk?" Pete asked, his eyes wide with bewilderment.

Clip laughed, "Oh, gosh, I forgot," he said. "That's what we call him—Monk—you know, Ted Monger."

"Oh, you mean the editor," Pete said. "I don't really know him very well . . ."

"Old Monk," he said, "you know I went to school with him, didn't you? Crazy old goofball—but he'll make an A-number-one editor. He was telling me some of his ideas about promotion on the train this morning. We both ride in from Stamford every morning. Like as not we're still both stoned from the night before. Sort of like the blind leading the blind. He's a great fellow, though," he said, "just the one to really put *Today* on the map."

"I thought that had been done," Jack Sheehan said quietly.

Pete looked at him and smiled; I coughed a little, and Clip West pretended it had never been said.

"About the sales meeting coming up in a few months," West said, "I guess that's one of the first things we should be thinking about."

"We had a really fine one last year. Out at Guerney's Inn. It really came off great."

"Guerney's Inn?" Clip looked blank.

"That's in Montauk." It was one of my few contributions to the conversation.

"We've been thinking of going to Greenbriar this year," Pete said. "I've checked it out and it seems fine. I got all the details."

"Monk and Tom were thinking of something looser," Clip announced quietly.

"Looser?" Pete said.

"Monk thought maybe Freeport."

"Freeport?" Pete's mouth was open again; I had the strongest impulse to hit him.

"The Bahamas," Clip said. "Tom half opted for Nassau, but they've got gambling in Freeport. That sort of clinched it."

Pete grinned weakly, so completely out of his depth now that there was hardly a trace of where he'd gone down.

"Oh—Tom's talked about it with you already?"

"We talked a little," Clip said. And then, "I do admire you running something with all those de-tails."

Pete said, half apologetically, "Well, there's always somebody that's got to make sure there are note pads and ashtrays—and enough chairs and light sockets—even gum and Life Savers. Somebody's got to watch out for those little things."

"*Exactly* what I told Tom," Clip said.

It was about the biggest put-down I had ever witnessed, but Pete didn't seem to grasp it at all. Or didn't let himself. We ordered our lunch right after that, each of us having another drink except Jack, who'd kept himself down to only half of his martini. I didn't even want to think of what *that* had cost him.

It was at the exact moment when I was draining the last of my first drink that I realized who it was Clip West reminded me of. Just add a full quota of personal charm and attractiveness, and you had another Tom Redford. But in his own way this one was a much more dangerous operator. I gathered from some of the things he had said that he'd probably come from a dirt-poor southern family. Unlike Redford, who worked out of impulse, this one thought things through. He was a planner. It also seemed to me that Clip West was someone who had probably *worked*—and maneuvered—very hard to get where he was. You got the impression that he had at some time in his life cold-bloodedly calculated that what he possessed were several valuable commercial virtues: being able to effect a quiet kind of ingenuous pleasantness, a willingness to work, and the ability to devote an unlimited concentration of effort and desire to anything that was important enough to him. I felt very strongly that he had added them all up and had come to the conclusion that—handled exactly right—they could get him just about any of the things in life that he wanted.

Pete Larsen was another story. I wondered at what age he had learned to apologize to everybody around him just for being alive. And then I remembered what it was like to work for Redford—how hard it was, sometimes, even for somebody as street-kid snotty as I was, to hold my own—and I decided that Pete Larsen was acting just the way he had to under the circumstances. Which didn't do a thing to lessen my feeling of wanting to take him by the shoulders and *shake*.

"Of course," Pete grinned sheepishly, "the best part of it for me is always at the end, when we show all the pieces—the presentations and posters and brochures—that the promotion department has turned out for the previous year. What I did last year was to string this clothesline all around the conference room—right from one corner to the next—and what I did was to use clothespins to actually hang all the promotion pieces onto the line. Boy, I tell you, it was pretty impressive."

"I was thinking maybe we'd use live models," Clip said.

"Models?"

"Oh, maybe just three or four. The way I figure it, these sales meetings shouldn't just be a rehash of what's been done the year before. I see it more like a"—his hands went out, so help me, as though he were shaping the creation of a new world—"a *projection.* You got to make it fast, exciting—something that'll make those space peddlers sit up and take notice. The way I figure it, each one of these girls—these models—represents a brand-new project that we're planning on turning out. Like, say, one is Miss Direct Mail, and we have this here cos-tume that she wears—like a bikini, maybe—only when you get up close, it's not a bikini, it's this fabric made out of pieces of direct mail. And we have another one for presentations, except we don't tell them guys any boring facts or statistics, we just have her dressed in, say, maybe a shift—only it's all made out of film strips. And maybe we get ourselves a new slogan for the magazine—something real bright and catchy——"

"Jim Canfield always shied away from slogans," Pete said. "He believed that, if you weren't careful, you could easily end up editing the magazine to fit the slogan."

"Tom was thinking maybe of something like 'The magazine for people who count.' "

"What does it mean?" I asked, but nobody bothered to answer.

"Anyway—listen, Pete," West was saying, "don't you get any ideas that any of this is all sewed up with Tom and Monk already. What I mean is—well, gosh, you've been around here all these years, ain't no little johnny-come-lately about to come in and reshuffle things right off-hand. You done a fine job—what I mean is, well, all those years, *Today* wouldn't have been the *same* if you hadn't been running that promotion department."

"Well, thank you," Pete said, his eyes shining like some grateful dog or something, "we tried to do our best," I almost gagged.

West reminded me of someone who would be unfaithful to his wife *and* his mistress. In this case probably some researcher at *Newsweek*, or *Life*.

"And you *did* it," Clip said, for all the world like some benign conqueror.

"Listen," West said, "I'm gonna need the help of every one of you if I'm not going to fall on my face in this new job." He actually looked around the table, from Pete to Jack to me. "The way I figure it," he said, "we got us the best little promotion group in this town—and, like I always say, in the long run it's not the product—it's the promotion.

"Not that *Today* ain't about the best magazine there is in the country," he added strongly. "Because, man, you can bet your last dollar it is. It's just that—well, you can't keep your light under a bushel. No matter how good it is, it's like I always believed and always will believe, 'It's not the product, it's the promotion.' O.K.?"

Jack finally finished what was left of his drink in one deep swallow; I didn't say or do anything; and Pete—naturally—said, "O.K."

"*Great*," Clip announced, grandly waving for the check. "Well, I guess we better get back to all that promoting work we get paid to do. I told my secretary to make sure she had about a dozen new note pads and a slew of sharpened pencils, because I want to get out memos—personal memos—to each of the salesmen in all the offices around the country. I want to get to know each one of them personally just as soon as I can—so I warned her to be ready for a bushel of work this afternoon."

"Who is your secretary?" Pete asked.

Clip West laughed, full of sham embarrassment, and said, "Oh, yeah, by the way, I meant to get around to that—I don't know where I'm keeping my brains today— Pete, Tom suggested maybe it would be a good idea if I took over Maryanne."

"Maryanne?" Pete repeated, as though he had never heard the name in his life.

"Your secretary," Clip said. "Tom figured it'd be a good idea for me to have somebody who knew the run of the place. He said he'd call personnel to have them send some new little girls in for you to interview. He said it wouldn't probably take more than a day or two till they found somebody. That's O.K., isn't it, Pete? I mean, it doesn't put you out—if it puts you out you let me know, you just let me know and we'll make us some other plans . . ."

"No," Pete said, "no, no, it'll be all right." And then after a moment, "Maryanne's a good secretary. She's worked for me for about ten years."

"That's it," Clip said, "like I said, if it should put you out, you say so. But Tom did think it'd be a good idea . . ." Then the check came, Clip made a big thing of signing for it, and we started on our way back to the office.

Walking next to me, Jack Sheehan said, "Like the poet said—you'd have to see it not to believe it."

I went directly to the ladies' room, and when I came back to my office, the shock of actually seeing Pete Larsen on the phone behind that third desk was enough to throw me. He was just hanging up as I entered. Jack was already back at his drawing-board, staring at the wall outside his window.

I went to my desk, not knowing whether I should tiptoe or do something completely outrageous.

Pete—his eyes wider and more dog-sad vulnerable than usual—finished hanging up the phone and said, "That was Mr. Wolffe. He wants you, Paula—and Jack—and three or four people from the other Promotion Departments who have never been to the main plant in Chicago—he wants you to plan on going there tomorrow. You're to be his guests at dinner in the evening and stay over and then take a tour of the plant the next morning."

"You have to go to Chicago with me."
"No, I can't."
"Yes, yes you will. I'll find a way."

And evidently he had done just that, he had found a way—even if it had to involve half a dozen other people as well.

"Did you say tomorrow?" Jack asked. And Pete said, "Yes, you're all to get up there in time for dinner. And then stay overnight. He said he'd made all the plans . . ."

And, angered, pleased, excited, frightened, I tried to put my finger on my one true reaction. And I couldn't.

"Well, it should be interesting," Jack said.

"Yes," I said, "that's what it should be—interesting."

At which point Jack went back to staring at his wall, and I at my typewriter. And as for Pete—we just pretended he wasn't in the room at all.

Just before I started banging out some nonsense on the keys, Clip West went past the door toward his office. For a second he stopped, pushed his head into the doorway and said, "I wanted to thank you for being my guests at luncheon. It was a real treat for me."

And, so help me, Pete actually said, "Thank *you.*" While I thought something that would have made pure and most perfect graffiti.

nineteen

AS SOON AS JACK AND I GOT ON THE PLANE I started thinking of the trip as *The Bobbsey Twins Visit Chicago*. The day was grey and cloudy and chilly, and for some reason, I kept thinking that my one hope of keeping my sanity through this whole thing was Jack Sheehan. I told him so, and he said, "I feel the same way about you. Let's both make a pact that we refuse to have this thing drive us absolutely crazy."

"Yes," I said, "I promise. We'll protect each other."

We actually shook hands—more solemn than we intended to be—and he said, "It's a deal."

Jack and I were fond of each other in a kind of brother-and-sister way, but, actually, there was something deeper than that. I think in a way we really did depend on each other. Considering the erratic pattern of my own actions, plus the fact that Jack was drunk half the time, our mutual dependence had its comic overtones.

I said, "Do you suppose we'll see our renowned chairman of the board?"

And Jack said, "Mr. Arthur Jefferson—and Miss Lo-rraine?" And we both laughed. It was just possible. We might survive the next day with our sanity, at that.

When we finally got to Chicago at about four in the afternoon, it was rainy and cold and we had to wait fifteen minutes for a taxi. Our instructions from Dave Wolffe, via Pete Larsen, had been to check in at the Regent Hotel and to meet in the lobby for dinner at seven o'clock.

As soon as we checked in Jack said, "I'm a little restless. I think I'm going to walk around for a while. Would you care to have a cup of coffee?" I said, "If you don't mind, I think I'd rather go right to my room." His hands were shaking as he lit a cigarette.

"No, that's O.K.," he said. "If there's anything you want just call me."

"Take care," I said—that, at the moment, being the sum total of my contribution to humanity, mankind, and Jack Sheehan in particular.

When I got to the room I hung up the few things I had brought with me. A thin chiffon blue-and-green dress that I would wear to dinner that night, and a red woolen suit for the tour of the printing plant the next morning. The hotel room gave me the same antiseptic creeps I always get from hotel rooms. The five wooden hangers hanging in the otherwise empty closet, the two drinking glasses in their thin sterile bags, the extra roll of toilet paper, the soft length of brown chamois cloth with the name of the hotel printed on it.

I read all the little table tent cards they always leave hanging around in hotels. One was about a restaurant in the hotel called Le Manoir, another folder told all about a snack bar. Another card told you what extensions to call if you needed: a physician, long distance, the valet service, and the front desk. I checked, and there was a Bible in the drawer of the night table next to the bed. Also a folder of hotel stationery and a blue ballpoint pen that had gone dry.

Having taken inventory, I filled the tub with very hot water and slid gratefully into it. I must have lazed there for at least twenty minutes; I could feel the chill coming out of my body like long, shivering threads. Finally, when I had soaked myself to a prune, I got out of the tub, dried myself off, and got into bed, pulling all the blankets, including the candlewick cover, up over my head.

I had no idea how long I had been asleep when the phone rang right next to the bed. I looked at my watch. It was six o'clock. I picked up the phone, expecting Jack Sheehan. It was Dave. He said, "Hi, I just checked in this minute; is everything all right?"

"What?" I said, and then, coming more than halfway awake, "Yes, yes, I'm fine. I was taking a nap."

"Oh, what time did you get here?"

"It was just around four, I guess."

"I have to make a few phone calls. Will I see you before?"

"In the lobby at seven, I guess." I felt vague and shadowy answering the question—cold and unreal and distant—wanting more than anything in the world to go back to sleep, with no decisions that had to be made.

"O.K.," he said, "we'll talk afterwards."

I hung up then, not thinking too much about what that was supposed to mean. I wanted to stay in bed, warm and safe, the whole night, and not be a part of any of it.

But there was no time. I made myself put my feet on the floor—that was the hardest part. Then I started to get dressed. In the middle of getting ready, I started to yawn—deeply, compulsively.

I fumbled with the catch at the neck of my dress, got mascara in my right eye, and managed to lipstick on a completely crooked mouth that I had to wipe off and start all over again with. In the middle of it all I couldn't stand how quiet everything was—except for my yawning—and I turned on the radio that was next to my bed. The first thing I got was a low, treacly tenor:

> "I'd like to get you
> On a slow boat to China
> All to myself alo-o-ne . . ."

I switched haphazardly.

"Rock of ages
Cleft for me
Let me hide myself
In the-e-ee . . ."

I switched off the radio.

At exactly two minutes to seven I left the room, closed the door behind me with a sneaky feeling that there was something I had forgotten to do, and walked down the green-carpeted hallway. I pressed the down button and studied myself in the glass panel that separated the banks of elevators. I looked O.K., I guessed. Surprisingly, there was very little you could see of either my reluctance or apprehension. Just before the elevator came I stuck my tongue out at the reflection in the mirror. I wanted to prove it was really *me* standing there. The fact that whoever was in the mirror stuck out her tongue, too, was some reassurance.

The elevator stopped and I got in. There were two women and three men in the car. One of them was Dave. Completely thrown, I said, "Hello."

He looked just as bewildered as I felt. Nobody in the car said anything else all the way down to the lobby. The ride seemed to take at least half an hour. I could feel Dave staring at the back of my neck. At last we reached the lobby and the door opened, and there, waiting directly next to the elevator, was the whole Kimberly contingent, all five of them, Jack Sheehan included. Jack said, "Hello, Paula—Mr. Wolffe." It was the sole indication I had that I hadn't become suddenly invisible. As for the promotion writers from *The Review*, after they got over whatever shock they might have suffered from seeing both Dave and me alight from the same elevator, I was afraid they would knock me down and trample me in their eagerness to "get to" Dave. There were three of them—all men—all around thirty-five, two of them with thick, dark hair, one of them going bald, and all of them imbued, obviously, with a passion to make points with Dave Wolffe. Steve Lunderman had come along, too. Unlike the others, he stood in the background and projected solidity. The only two people in the whole group who gave any impression of being human beings were Jack Sheehan and Dave. Jack was quiet, but not so much that one got the impression of deliberate quietness. He was merely there. Come to think of it, I'm not really sure that Jack would have known what the expression "making points" meant.

The real revelation, though, was Dave. He seemed comfortable, confident, and relaxed. Which was even more remarkable when you consider the little globs of adulation forming around him. "Oh, Dave, about that great promotion idea you had," and "Dave, I want to tell you about this thought that I got after our last conversation," and "Oh, Dave, this," and "Oh, by the way, Dave . . ."

It was Jack Sheehan who, smiling, very quietly said, "Well, Mr. Wolffe, do you suggest we all pool our money and get rich in Alaska?" And, just as quietly, Dave, grinning, said, "Not unless it's money you can afford to lose." At which the three men from the other magazine all looked surprised and taken aback for a moment, more so as Dave said, "Nothing's been verified as yet. When I'm ready to put my own money into it, I'll let you know."

The restaurant was right in the hotel and again it was just a matter of luck that I didn't get trampled by the three writers jockeying to get next to him. As it was, I ended up across the table from Dave and next to Steve Lunderman, who I don't think said more than five sentences during the whole meal, each of which was prefaced with either, "When I worked for Howard Hughes," or "As my buddy, Tom Redford, would say . . ."

After a while I began to sort out each of the three writers. The one who was going bald had a way of prefacing every statement with "In my opinion . . ." "In my opinion, Dave, I think *The Review* is experiencing a true cultural renaissance." Once, a little later in the evening, he even ventured so far as to say, "In my opinion, the food in this restaurant is very excellent."

The one who had glasses kept asking Dave for *his* opinions on things—everything from Berkely to the Common Market—from the state of the theater in America to views on cigarette advertising. "I think without any doubt there's validity to the statement that enough smoking will kill you," Dave said. At which point he lit another cigarette and inhaled. And grinned. "But so will a lot of other things."

The third one, with thick dark hair, played it the "family route." He gave the names, nicknames, birthdays, and habits of each of his four children, and he prefaced just about everything he said with, "Speaking as a father . . ."

The only person who seemed completely himself was Dave. He said, "Did you hear about the Kimberly flag?" No one had, so he said, "Well, as you know, we're taking over two more floors in the building. When that happens we'll be the major occupants. Anyway, the whole thing has involved a long series of meetings and more meetings. Last Wednesday night there must have been a whole barrage of lawyers involved in this meeting. Most of them representing the building, naturally. It went on and on and on. Finally, it's about four o'clock in the morning and we're finished. Everybody's beat, the room is full of smoke, hardly anybody can even talk anymore—much less argue. So just then, when the lawyers are packing their briefcases and the last papers are about to be signed, I say 'What about the flag?' " He grinned like a kid; you could see how much he enjoyed telling the story. He went on. "Everybody looks at me because so far there hasn't been any discussion about any flag. The building has two flagpoles, right? One's the American flag. The other one's Inter-Com Brands, since they always rented more space than anybody else, before we took over. I look at them and I say very quietly, 'I've got to have a flagpole for the Kimberly flag.' " He raised his eyebrows. "Of course, we don't even *have* a flag as yet. But that's a minor point. Anyway, I say, 'The Kimberly flag has to fly out there every day or else it's no deal.' And that starts the lawyers sputtering—Inter-Com's always had its flag out there. It's important to them. Especially when they have brass coming from out of town to visit the office. They've *got* to have a flag, et cetera, et cetera, et cetera.

"This nutty discussion went on for three hours. Would you believe it? At seven o'clock in the morning, we reach an agreement. It's simple. Kimberly's flag flies outside that building every day." He paused for effect.

You nutty *actor*, I thought.

"What about Inter-Com? Well, it goes like this: we agree to have the Inter-Com flag on the second flagpole out there whenever they've got VIP's coming in to impress. But when that happens—it's the American flag that's lowered. Under no circumstances does the Kimberly flag ever come down."

He laughed. The street kid who had conned the Establishment. It was childish—immature—unprofessional. And yet, the way he told it, you couldn't help yourself, you were on his side; you had to enjoy it too.

And then someone mentioned Brad Rusk, and Dave lost the warm and witty and natural attitude he had had all evening. He said, "It's men like Brad who make publishing a significant profession," and I thought: Come *off* it—where are you? You can't believe any of this—who are you trying to kid?

But he went on, "Editors like Rusk are born once in a generation." And after a while I had to admit that he was really saying it, and it was all a complete contradiction. It was only when he spoke of Brad Rusk that his viewpoint seemed distorted, out of perspective. I found myself shivering—not only because this was no person I knew, but because it was somebody else completely who had told me about his childhood. "*I have this dream . . . My father is dead.*" The thing that was much more deadly to realize was that in all the conversations we had had, never once had I even said a word to try to dissuade him from this obviously deadly alliance. At least for the others, the battle lines were drawn—Moshier and his crew on one side, Brad Rusk and Dave Wolffe on the other. And then there were people like me. Good old Paula, *everybody*'s friend—which meant what? *Any*body's? *Whose?*

It was almost ten o'clock when we all finished dinner and drifted back toward the elevator. For a moment I felt awkward, unsure, in some way exposed. It was Dave who broke the awkwardness, who turned to us and said, "I want to thank you very much for coming. I may not get a chance to see you tomorrow. If not, I hope you enjoy the tour, and I'll see you back in New York." And with that he sort of waved and got into the elevator.

For a moment after he left, the five of us stood there stiff and tongue-tied. Then Steve Lunderman, seeming to come to life, said, "Can I buy anybody a brandy?" At which point the three men from *The Review* sort of looked at each other as though each one was waiting for the others to come to a group decision.

I said, "Thanks, but I think I'm going to go right up," and Lunderman said, "Well, shall we all meet down here tomorrow morning? Say about eight-thirty?" I said, "Fine. Good-night," and pressed the button for the next elevator going up.

The room seemed empty the way only a hotel room can. It was raining hard now; you could hear it against the window. I turned on one of the lamps, put my key down on the glass-topped dresser, stepped out of my shoes and then just stood there, absolutely motionless, until the phone rang.

Dave said, "I've got to see you. Can I see you?"

"All right. For a little while."

"What room are you in?"

"Four twenty-three."

"I'll be right there."

I took off the blue-green dress that I had worn to dinner and put on the yellow dressing gown I had with me. Then I took off the dressing gown and put the blue-green dress back on.

There was a knock at the door. He had brought a record with him, and a small record player.

"I wanted you to hear the 'Schelomo,' " he said. "It's the piece I was telling you about." He opened the record player and set it on the bed and started the record.

It was like nothing I had ever heard before. "It's so desolate," I said. "It's like being so lonely you can't stand it anymore."

"Except if you keep listening you know that you can."

"Yes."

When it was finished we sat there in silence, he on the floor next to the bed with the record player, I in the striped yellow-and-green chair across the room.

Almost as though he knew it was what I wanted him to do, he started the record over. I felt as though I could go on listening to the music for a long time. I felt weightless, bodiless, completely at one with the music and with the person I was hearing it with.

Halfway through the second time Dave stood up.

"I have to go," he said. "You keep the record player. I bought it for you."

"Dave—where are you going? What's the matter?"

He stopped, already halfway to the door. "It's all too pat," he said. "Almost a joke." It was as though he couldn't bring himself to look at me.

"I don't know what you mean."

And then he looked at me, a look I will never forget, a look that went past your eyes into your brain, and past your brain into something else.

"You knew why I set up this whole thing. A tour of the plant. Everybody from promotion who had never been here. And I would give a dinner for them *the night before.* We both knew why, didn't we? There was never any doubt about it—was there?"

And I said, "No."

He was at the door. I could barely hear him. "I'm sorry," he said. "In addition to everything else, it was a *cheap* maneuver. I guess it's the only kind I know."

"Dave—Dave—you don't have to go right away. Finish the record."

The music had reached a crescendo now. It poured like tears, or like the sun melting ice and the ice becoming tears—hot, raw, scalding.

He came back and, in silence, we listened to the music to its end.

And then, after it had been quiet for a while, I asked, quietly, "What made you change your mind?"

"I don't know—I think just seeing you at that table tonight. Not like anything I'd ever known. And all of a sudden I was sick with everything I had done up to that minute. All the ploys I had used—anything to impress you, to confuse you. Today—tonight—bringing you all here—making up a story, a reason—knowing that no one could question that reason. Just to get you here, because you wouldn't come otherwise, because I knew you meant it. All of a sudden I saw how cheap the whole thing was. And do you know what made me see it clearest of all? Because I trusted you. That's what I knew tonight sitting there—that all my life I'd always liked people, but after I'd *checked.* My family, those people at the table tonight—I represented power, someone who could maybe do them some good; it created a working arrangement. But with us—with you—it was different. I trusted you—without checking, without anything."

"Dave," I said, suddenly afraid, much more afraid than I had been before. But he went on, "Do you think I don't know the image I give—Seventh Avenue opportunist . . . unscrupulous, ruthless? There are millions of us. Do you think we don't know it?"

I wanted to stop him then, to tell him that I was one of those, that he was mistaken, but instead he went on.

"Don't fool yourself. We know ourselves. And most of the time we capitalize on it. If I could scare them shitless at Kimberly by seeming to be a hatchet-man, a liquidator, an unscrupulous manipulator—if it served my ends, of course I used it. Rule by fear—it was practically a motto for me.

"Until you," he said. "All of a sudden there was somebody I didn't want to be afraid of me. I mean really not afraid. And you seemed that way. Oh, I wasn't ever *sure*—but you seemed that way. And at times I used to almost be afraid myself. At some point you would weaken—you would change . . .

"And then tonight, I *really* knew it. I knew that it wasn't an act, that you really weren't afraid of me. And that was when I realized I trusted you. Do you have any idea how long it's been since I trusted anyone? It was my mother who taught me how *not* to trust—how that was for *schlemiels*, dumb *schlemiels*—failures. To distrust was to be a man."

"Dave." I tried to stop him, but it was as if he didn't even know I was there.

"Even when I was a kid, I *knew* they thought I knew. I looked as though I knew—it was the same thing. And it never changed—afterwards.

"You can't imagine how easy it is to convince people that you know everything, that there isn't anything you can't do—because that's what they're looking for—someone who knows. That's all they want—and smart Sephardic Dave Wolffe—even the name was right—he knew. Others might guess, or make mistakes, or have to learn, but he *knew*. Sometimes I got so sick—and lonely—of the lie, but it was the one I had chosen. The one that fit.

"And then, all of a sudden, the only person I ever saw—you—you in a whole roomful of people—that morning when we introduced Monger—you *doubted*.

"Do you know what that meant? Somebody I didn't have to

keep proving I *knew* something to all the time? It was the strongest possible attraction any woman could have had. And it was you." Then he stopped, and I tried to think of what it was I had wanted to say, but I was quiet too.

"I better go now," he said. "You're right. All the times you told me we shouldn't meet, all the times I wouldn't let you say it, you were right. I'm sorry I gave you such a hard time. I'll leave you alone now."

"Dave . . ."

I raised my arm to reach him—and I touched him, and he seemed to be aware of it before I was, and compulsively, almost automatically, our arms were around each other, and it was as though we both knew what neither of us had known before: how much both of us were starved for the touch of somebody, for this touching which was the symbol of all touching.

For a second I could feel his surprise, almost his rejection of this motion of need, but I was not surprised; for me the reality of this time had existed all the way from the beginning.

I felt his fingers probe for, find, and then pull down on the zipper of my dress . . . inch by inch—conscious of the silver teeth of the zipper parting, opening. I felt my flesh reaching out toward touch—straining, searching for the feel of something against my flesh. Him.

"Paula . . ."

Undeniably—almost against his will, but undeniably—his arms loosened only to draw me more tightly to him.

"Dave—I need you too, Dave—I didn't know—Dave . . ."

And that was the truth of it. It was not *his* first need we were willing to be, but my own. I felt as though I were the first person in the world for him—his family, his brother, his wife, his mother, everybody who had not been what he thought I was.

His face pressed against my hair; he said, "Oh, Paula—I want to get you so many things. The pearls. And a white seal coat. And I want us to go to the Greek Islands . . ."

"Shhh . . ."

"I want to give you so many things. Things I've always dreamed about. It'll be so wonderful."

"Shh—David . . ." I said, pressed against him, half laughing, spilling over with joy and immensity. "David—David—you don't have to talk . . ."

And very quietly, holding me against anything that would ever come to interfere, he said, "I want to—please, I want to— Don't cheat me . . ."

I held him, not understanding, not thinking at all . . .

I touched his hair. Hair for which I had imagined dislike. Dark brown, dry-looking—foreign to me. And instead . . . the touch of it—the smell of limes—sudden—quickening me.

And his skin. That I imagined cold. I brushed my fingers—my face—against the warm softness of it. Softness of the delicate skin beneath his eyes. Warm softness of his throat. Strong smoothness of his forearms. Dense haired smoothness of his chest. Arms. Legs. A small, deep scar across the right shoulder. Kissing it. Scar that had been gotten where? . . . somewhere . . . Feeling for a moment outside. Unknown terrain of this strange body. Existing beyond me. Layer upon layer of wonder. Warm skin of life I had never known.

My hands moved over his body. Face. Throat. Chest. Belly. Thrusting spire of flesh. Moving. Growing. Red-crowned flower thrusting out of grass-hair. And down. Beneath. Strange, terrible. Double-pouched wonder of seed. I could feel the air against my naked body, making my body feel lonely, longing, yearning for the touch of him that finally came. His hands were like something I had always known. And yet there was strangeness. Sharp. Pointed. Strangeness of strange hotel room. City I did not know. Bed that was not my bed. Strong rain against the strange windows. The room lighted. But dim. Deep corners of shadow. Hands touching me. Beneath. Above. Under. Touching. Rounding. Cupping. Caressing.

And his excitement like a child's. Each baring of myself—clothes strewn on the floor, on the bed—each loving of him was as though he had never given himself before. And as for me, I did not think; I didn't dare think.

It was as though I had never really experienced the body of a man before. As though I had never appreciated the power of it. Or the urgency. Or the vulnerability. His tongue explored my body. My body . . . open . . . billowy . . . soft there . . . likes waves parting . . . and so open . . . so open . . . my God . . . what would happen now? . . . until shorn of shame . . . waiting . . . wanting to give . . . give . . . to give what was given . . . His tongue exploring my body. And mine his. Our hungers welded us to each other. Faster. Together. "God—my God!" Was the cry mine? . . . his? . . . torn, ripped cry of wonder—of sensation—of overwhelming frenzy.

"God!" Cry that was mine . . . cry that was torn . . . ripped . . . cry that welled into screaming. And our bodies wild—wild. Thrashing . . . churning. Our hungers welding, feeding, until they rose higher and higher and there was no control now, no stopping, only a plunging deeper and deeper to the one final and unanswerable end.

Awakening, I smelled the smell of limes and lay still, feeling nothing—and at the same time all-feeling. He was asleep beside me.

It was a mortal sin—if I died right now I would go straight to hell. That's what it had said in the catechism. For a moment I felt sick with fear. And then—almost by itself—my hand went out and rested on the smooth roundness of his left buttock. He was asleep. How quiet he looked, how naked, how vulnerable. For a moment I wanted to wrap my arms around him, almost as though he were a child, a little baby. My hand moved up his back and across his shoulder. In his sleep he muttered a little moaning sound, the lonesome moaning sound of sleep and, not wanting to wake him, I still could not keep myself from putting my lips to his. Gently, so I would not waken him. His lips parted a little; I could feel his breath coming into my mouth . . .

Suddenly the phone rang. For a moment I doubted its reality. It was as though it—the phone, my being here, Dave—were all a part of something I had imagined. But then the phone rang again. Slowly I picked up the receiver.

A thin, ubiquitous voice said, "Miss Jericho?"

"Yes."

"Oh . . . well, I'm sorry to disturb you, Miss Jericho, but—well, we have an emergency call for Mr. Wolffe . . . from New York . . . and, I mean—well, Mr. Wolffe's number doesn't answer and, well—I'm terribly sorry to disturb you like this but, you see, we—we're calling each of Mr. Wolffe's guests to see if perhaps any of them may know where Mr. Wolffe may be . . . and . . . well . . ."

I didn't answer him. Instead, seeing that he had awakened now, I handed the phone to Dave. "It's for you," I said.

He was only on the phone for a minute. He said, "Yes," and then, "Yes, I'll be there." Then he hung up and quickly started to get dressed.

"It was Rusk," he said. "There's been an emergency. I have to get back to New York right away."

Very quickly he started to get dressed.

"Go back to sleep," he said. "It's only one o'clock." When he was ready to go, he came over to the bed and pulled the covers up tight over me. Then he bent and kissed me very gently on the mouth.

"Goodbye," he said. "Miss me." Then he was gone.

twenty

JACK SHEEHAN AND I WERE THE FIRST two to meet in the lobby the next morning. His face was white; the skin seemed stretched over his bones, and I noticed how carefully he kept both his hands pushed into his pockets where you couldn't see how much they were shaking. We went into the cafeteria to have breakfast, and it was fine, except when he had to lift the coffee cup from its saucer to his mouth. It was then, for the first time, that he referred to the subject we had always been so careful to avoid.

"Paula," he said, "I don't believe I've ever told you how I came by being a drunk, have I?"

But there was no time for an answer, because just at that moment Lunderman and the three little elves from *The Review* arrived, all bright and shiny like the painted-up fronts of buildings.

Lunderman just sat, sort of grey around the eyes, looking straight ahead. The three little writers all in a row studied their menus with great and serious deliberation, finally coming up with three identically brilliant decisions—"Orange juice large, two eggs sunny-side-up, and black coffee."

As for me, I pretended that I had had some breakfast in my room before coming down. The idea of eating was pure fantasy.

No one referred to Dave. If the desk had, as they said, called everyone's room number—"all of Mr. Wolffe's guests"—then evidently no one was about to be the first to mention it. Or maybe it hadn't even happened that way—maybe they had known at the desk exactly where to reach him . . .

"*What room are you in?*"

"*Four twenty-three.*"

I wondered if I would see him today.

"Well," Steve Lunderman said, *à la* games mistress, "are we all about ready?" And we all were.

We were to take a cab from the hotel over to the executive offices on Michigan Avenue, and from there a delegate from Kimberly would take us on to the plant and *le grand tour.*

When we got to the Kimberly executive offices, there were two men waiting in the lobby to receive us—one a G. D. Davison with a tight, dried-out-prune kind of face, and a leering expression that went from ear to ear, so deep and so constant that you never did get to see what his eyes looked like. The other one's name escaped me—as did our reason for stopping here at the building, although I guess the Kimberly lobby was something I would hate to have gone through life and *missed*. The entire back wall from floor to ceiling, which had to be at least twenty feet, was painted in a design of huge trellises, around which twisted every possible size and color of rose you could imagine. Plus some that went considerably past imagination. Near the top, the trellises just sort of smeared off into a big globby blue mass that was supposed to be the sky, clotted with a lot of fat clouds. And all around this, attached to the wall or hung from the ceiling or fastened to the floor were huge wire cages full of birds. There were green birds and pink birds and blue birds and some that were black. They were big and small and medium-sized. There must have been at least a hundred of them. And they had only one thing in common. They were all making noise at the same time.

G. D. motioned behind him with the pride of workmanship of a man who has done it all single-handed and said, "I knew Mr. Wolffe wouldn't want you to have missed *this*." And we all smiled and allowed as how we were sure that was so.

For ten minutes we went through all the awkward saying of things that people say who have nothing to say to each other. Then it turned out that what we were waiting for was the appearance of a "Mr. Folsom from Circulation," who was to actually conduct us over our tour of the plant. He arrived finally, half on the run, panting, sputtering excuses for having kept us waiting. He was a small, pale-faced man with thin ridges of white hair through which you could see the baby pink of his scalp. Everybody was introduced all around and, finally, waving goodbye to G. D. and the other man, whoever he was, we all got into a big black car, with Mr. Folsom driving. The Chicago printing plant which operated in tandem with the one in Dayton was about forty minutes from the Kimberly building, and Folsom kept up a running commentary all the way. "As far as publishing techniques go, Kimberly has a reputation for—"

"It's like being so lonely you can't stand it anymore."
"Except if you keep listening you know that you can."

The printing plant was a big grim slab of a building that for some reason immediately reminded me of an orphanage. And we tramped around inside of it for two and a half solid hours. The one overall and inescapable lesson I learned from it was that I should have worn flat heels. If I wasn't slipping around in oil smears I was climbing high metal steps or stumbling over snaky cables. More than once I would have walked directly into a machine or somebody operating a machine if Jack Sheehan hadn't steered me away just in time.

"We'll keep each other sane."
"I promise."

I was only vaguely aware of the facts and figures that Folsom kept sprinkling through the tour. All I really registered was that practically everything he said appeared to be in direct contradiction to what I could see going on. For instance, just as he was expounding on the fact that "this complex of Kimberly machinery you see all around you represents the finest, most up-to-date mass printing facility in the world," I turned around to see a whole row of little old grey-haired ladies bending to lift stacks of magazines from one place to another. God knows, I was hardly an authority on what made or did not make an up-to-date printing plant, but when we came up to three huge machines, all of which were "closed for temporary repairs," when we saw whole slews of what obviously could be machine operations actually being done by hand—usually little old ladies' hands—and when for a moment huge waves of paper began springing out all over the place like wild ribbons of white shaving cream, I had my strong, if unauthoritative, doubts.

We went back in a company car to the Kimberly Building where we were to have lunch in the executive dining room.

"There are millions of us. Do you think we don't know it?"

G. D. was our official host at lunch and, deferring to authority, the up-to-now loquacious Folsom became as quiet as a totem pole.

G. D. was evidently the resident raconteur. His stories were old, usually dirty, and always long-winded.

At one point, in the middle of a particularly boring accolade about Kimberly and all the people thereof, he said, "It's kind of like this man whose wife is berating him for the fact that he's fallen well behind in performing his—well, his—eh, husbandly functions." His eyes bounced in sticky glee. "And the man looks completely taken aback, and his eyes widen into the perfect picture of innocence and amazement, and he says, 'But honey—you don't understand—they told me if you gave at the office you didn't have to give at home.' "

"I trust you. Do you have any idea how long it's been since I trusted anyone?"

Almost everybody at the table laughed completely on cue. The three writers from *The Review,* all sitting side by side, laughed a good forty seconds after everybody—even Folsom—had stopped. Encouraged by this reception, G. D. went on through his entire repertoire of barnyard humor, laughing throughout—his own best audience.

After almost every one of these earthy jokes he would bend down the table at me and say, "My apologies to the ladies present," and leer. But—outside of the apology—my presence didn't seem to hamper his storytelling very much.

At one point the conversation got around to *Today* and G. D. said pontifically, "She's what we call our Tiffany product. 'The uncommon magazine edited by uncommon people.' " I had never heard that one before; I wondered whether it was the "jim-dandy" slogan Tom had been searching for, or whether it was just something G. D. had made up out of his very own head.

"Not that I couldn't do with a little less of some of those long text pieces and a few more pictures of people," G. D. continued. "You know what I mean, people enjoying themselves in the sunshine. Like that story we had last year about St. Tropez (Saint *Tro*-peez) . . . Sure, I was interested about the scenery there and the gambling and the politics and all that. But there weren't enough *people* in it—you know, like I said, all that sunshine and everything."

"When you say people, do you mean *girls?*"

It wasn't so much what I had said, which was pretty tame, really, and actually just a question, but the fact that I had said *anything*—that I had been brazen enough to question anything . . .

For a second G. D.'s little cookie-mouth pursed into the shape of a cherry, and his eyes were like two little dots of surprise. The three writers from *The Review* all sat and gaped at me, their forks frozen in mid-air. Steve Lunderman stared down at his plate as though trying to divorce himself from any responsibility for even being there.

We must have held that charming tableau for at least a full minute, until G. D. blinked his little eyes four times in quick succession and laughed. "Well, now, Miss Paula—it sure is fine hearing a woman speak up and express her opinions right out like that." He said it, but you could tell he didn't mean it.

Nobody said anything at all for at least a minute. It was so quiet you could even hear people chewing and swallowing.

"Paula . . ."

"Dave—I need you too Dave—I didn't know . . ."

Pushing his wizened little face back into its usual smiling dried-out-prune pattern, G. D. said, "Anyway, I guess neither of us has anything to fret about. Ted Monger's got some real brilliant ideas. He'll make something great out of that magazine in no time."

Somebody said, "I thought it *was*." It was Jack Sheehan who had spoken.

Again the tableau froze into pattern almost immediately. For a moment I felt as though I were Jack—like a fly pinned into place by a half-dozen sets of sticky-glue eyes.

"Well now—that's so," G. D. said slowly after a while, threading out every word. "You're pretty close to the fact, young man—not a hundred percent, but pretty close. What I mean is—well, it's like a fine old lady beginning to show her age . . . Doesn't make her any less beautiful, in a manner of speaking—it's just that—well, she's a little out of step with the times, not as peppy as she used to be. What she needs is a shot in the arm—a little goosing up—if you'll forgive the direct language, Miss Paula. That's all the old girl needs, and she'll be as frisky and full of beans as you could want . . ."

I didn't say anything, but I must have shown my reaction. At first G. D. didn't respond at all, then he decided on a low, long chuckle that was positively spooky. As he chuckled, he rocked from one buttock to the other, while his tiny eyes glinted in amazement. "Why, I'm surprised, Miss Paula," he said, "a charming young lady like yourself being against updating and improving things."

In the back of my head, as clear as anything, I could hear Tom Redford saying, "The trouble with you is you don't know your place, is all . . ." I figured if he saw me now he'd have to change his mind. If I *did* have an opinion, all I gave to show for it was a sickly smile.

"People have got to stay flexible," G. D. said. "Why look at an old codger like me. I still get pretty juicy when I see some of them new see-through thing-a-ma-jigs on them sassy little girls sashaying around. Now you take some people, they been putting them new fashions down—say it's unladylike, putting ideas in men's heads and stuff like that . . ." He shook his head, enjoying himself completely now. "As for me, I say them ideas always been there anyway. So what if you put a few pretty pictures in the magazine—ain't no pretty girl's face ever *hindered* the selling of a magazine, I can tell you that for sure."

I'm not *sure*—I guess I never will be, but I think I might have said it *this* time, except once again Jack Sheehan beat me to it.

"Are you sure it's faces you're talking about?" he said.

One of the *Review* writers coughed into his soup, the other two went through similar, if less noisy, paroxysms.

"Why, what do you mean exactly, Mr.—eh, Mr.——"

"Sheehan," Jack said. "Where I get the idea is from the dummies of the next three issues. The first three issues being edited by Mr. Monger. According to my rough count there are approximately twenty-four navel shots in the first one alone."

"Naval shots?" So help me, even when G. D.'s mouth fell all the way down to his chin in amazement and confusion it still looked as though he were smiling. "Naval shots?"

"Navel," Jack said quietly. "Girls. Not ships."

Steve Lunderman coughed. The writers stared at Jack as if he had dropped from the sky.

"Tch'tch . . ." G. D. chided him, "you sure are strong-minded, Mr.—eh—Sheehan. Like I said, a few pictures of pretty girls in a magazine never hurt the circulation any."

"Five naked Arabian women bathing out of a makeshift tub . . . a story on Israel that bypasses the subject of its friction with the Arabs but concentrates on 'The High Rate of Illegitimate Births Within the Kibbutz . . .' " Jack's voice was so quiet you could hear Lunderman's cup hit the saucer as he put it down.

Finally G. D. smiled, all teeth and little eyes. "One thing—one little thing may have slipped your mind, young man. It's the editor who decides *what* goes into *Today* and what doesn't. Right?"

"Right," Jack said quietly. "Exactly right."

"Then you got nothing to be afraid of. Because Ted Monger's got some great ideas. Real peppy ones too. He's even thinking of putting in a whole twelve-page fashion section in each issue. Now for instance, Miss Paula, you can't tell me a lovely young lady like yourself ain't interested in *fashion*." He turned and smiled at me his slimy best.

"I love it," I said. "In *Vogue* and *Harper's Bazzar* and *Mademoiselle*." I did manage to get that much in. Big deal, as my mother would say.

At that point the waiter decided that this was as good a time as any to risk it, and he started to take away the plates.

For a little while the conversation strayed elsewhere. But I guess Jack must have struck a nerve with G. D., because just as dessert, a sticky lemon-meringue pie, was being served, he said, "Now you take this Orin Kreedel as an example."

Immediately I could feel every hair on the back of my neck prickling. G. D. said, "No doubt he was a fine editor. Wouldn't anybody dispute that. Just didn't have the knack of getting along with people . . ."

"Excuse me?" Jack said.

"Too rigid—too uncompromising—didn't know how to give a little here and there. One of the first rules of business, you got to learn how to give a little here and there if you're going to get along with people."

"*But*—" But even before the word was out of my mouth, Jack had interrupted, had stopped me, giving me no chance to continue. Polite, gentlemanly, quiet Jack. It was almost as though he were running interference for me, keeping me from a flying jump off the roof.

"But wasn't it Orin Kreedel who was mainly responsible for all the great writers coming into the magazine to begin with?" There was no stopping Jack now. I didn't even try.

"Hemingway and Orwell and Faulkner and Steinbeck and—well, you name them. Even when *Today* couldn't pay half as much as they could get elsewhere. Because they wanted to be in *Today*. First of all, because of what the magazine was, as Jim Canfield had made it. And secondly, because Orin Kreedel is, I expect, the kind of an editor that a writer prays for."

It was wild. As though he couldn't stop himself. But it wasn't as if he were saying it for himself. It was for me. As though he were saying these things to keep me from saying them—as though he knew absolutely, infallibly, that I would have said them if he hadn't.

"One last thing," Jack said to G. D. "I'd like to ask a question. Considering all the aspects—who do you think the better editor for *Today* would be? Ted Monger—or Orin Kreedel?"

G. D. laughed. It sounded a little hysterical. Then he said, "If you were as old as I am, young man, you'd know that questions can't be answered as simply as all that."

"Why not?" Jack said quietly.

"Well . . ." G. D. sputtered, "because—because there are a lot of other things involved. I had a few meetings with Orin Kreedel before this whole thing blew up. All I said to him was to co-operate with Ted Monger—that's all. And you know what he said? He said he couldn't. He said it was a matter of principle.

"Can you imagine that?" he sputtered. "A whole lifetime as an editor. Twenty-three years on *Today* alone. One of the most respected and well-known and best-paid editors in the country. And he threw it all away. And for what? For a *principle*—" He stopped, the incredulity in his voice so absolute and complete there was nothing else to say.

"What else is there?" I said it quickly, almost without a breath, positive that if I had hesitated a second, Jack would have said the same words—again shielding me from having said them.

Nobody moved. G. D. didn't even remember to smile. I stared at G. D., waiting for an answer. It took a while, but I finally got one. What G. D. finally said was, "Well, it was a real pleasure having you folks visit with us like this today. I have Dave Wolffe to thank for having the idea in the first place." There was no mistaking the dismissal of the group in the way G. D. put down his napkin, scraped back his chair and got to his feet.

It was amazing how fast the whole thing happened after that. Before I knew it we were out of the dining room and down into the lobby. I will never forget the clatter of all those birds among the roses. It was almost like a record playing at the wrong speed.

"Well, it's been real interesting having all of you," said G. D.

We stood there, shifting from one foot to the other.

"Real interesting . . ." G. D. said again, and smiled his shifty little smile.

Steve Lunderman announced importantly that he was going back to the plant to see about some last-minute positioning for *The Review*, and expansively asked all of us if we cared to go along. The three writers immediately jumped at the opportunity. I said—trying to keep the desperation out of my voice—"I think I better get back to New York."

Jack Sheehan said quietly, "Thanks, Steve, but I think I'll be getting back, too."

It was amazing how relatively calm we both managed to sound, considering that between us you could collect enough tension to blow Chicago right off the map.

Once we got onto the plane, we both sat there sort of mute with relief. Five minutes after the plane started, Jack said, "Thank you for doing it. I'm very grateful."

"What for?"

"You know—what we promised. To stay sane. I'd never have made it without you."

"I'd never have made it without *you*." Realizing suddenly that that was the truth.

Once, after we had been quiet for a long time, I tried to say something about what he had done for me at that ghostie luncheon—going out of his way to keep me from cutting my throat, very probably at the expense of his own.

"With Redford being my 'protector,'" he interrupted "my job's just a matter of time. You seemed upset. I thought you might say something you didn't really want to."

Almost the entire rest of the way home neither of us said anything else. For the moment I escaped into almost completely numbed tiredness. I closed my eyes and tried to sleep. But every time I opened them a wave of raw and urgent sensuality would wash over me.

"*Goodbye. Miss me . . .*"

"Well, I guess the trip was worth it," Jack said suddenly.

"Why?"

"For one reason." He smiled wanly. "You know those birds in the lobby? I asked the receptionist who fed them. They have a man who comes in every day from the outside."

"Just to feed the birds?"

"Yep."

"Well, I guess that's the whole story of Kimberly in a nutshell."

We laughed. Weakly.
We were almost home. We had made it. Anyway, almost.
The plane touched down smoothly at the airport.

"*Miss me.*"
"*Miss me.*"

twenty-one

I HAD MEANT TO GO RIGHT HOME ONCE THE plane landed. It was already late afternoon. But for some reason I went back to the office. When I got to my desk I thumbed quickly through the telephone messages but—if I was looking for something special—I didn't find it.

Feeling restless, I drifted down to Tom Redford's office. He wasn't there. Jane Perry looked up from her typing.

"Oh, hi, Paula," she said. "Did you have a good time?"

"Oh . . . yes."

"Where's Tom?" I said.

"Home, I guess."

"Isn't he feeling well? He seemed all right when Jack and I left yesterday."

"That's when it happened. Just a little while after you both left." She looked at me—calm, quiet—gauging me, as I felt she did a lot of the time, weighing her loyalty to Redford against the pity or friendship or whatever it was that I think made her worry about what might happen to me in this whole mess.

Finally she said, "I'm supposed to say that he's home sick. Actually I think he was told by Moshier to absent himself until told to return."

"But why?"

"Who knows?" And then, unusual for her who usually kept things somewhat distant and impersonal, "Are you all right, Paula?"

"Sure," I said defensively. "Why?"

"I don't know. You just look sort of—*different*."

I laughed, or something like a laugh. "That's what a day in Chicago can do for you."

I started to say something else, but just then I heard a phone ring down the hallway. I must have gasped because Jane looked up and said, "What's the matter? It's probably just one of the salesmen's phones . . ."

But I didn't wait to hear the rest of it because I was running to my office, my heart pounding, positive that it was my phone and that it would stop ringing before I got to it.

I half tripped over the telephone cord and grabbed at the receiver.

"Hello?"

"So did they give you so much culture in Chicago you forgot how to use the telephone?"

The disappointment scraped like a splintered chicken bone in my throat.

"Oh, hi, Vivian. I just got back. I was going to call you."

"Are you coming home soon? Should I start dinner?"

"Oh—well . . ." It occurred to me that I didn't know *what* I was going to do. "Maybe you better go ahead without me."

"O.K." A pause. "You're sure you're all right."

"Yes, I'm all right. People don't die just because they go to Chicago, you know." I hadn't intended the spill-over; on the other hand I hadn't been able to stop it either. "I'm sorry," I said. "I'll see you in a while—O.K.?"

"O.K. Anyway, it's salmon. It doesn't get cold."

"Fine. I'll see you."

"Yes," she said. "Sure."

After I'd hung up the phone the whole place seemed much quieter and emptier than it had before.

Almost in a panic about being left alone—not that I understood why—I hurried back to Jane Perry's office. She had left; the whole place seemed empty. I rushed back to my office, hurrying like some nut to scramble into my coat and pick up my things, almost as though I were as urgently compelled now to get out of here as I had been to come back.

It was just before six, and the streets were crowded. For some reason I couldn't stand the idea of people packed around me in the subway or a bus, and I didn't have enough money for a taxi.

I decided to walk home. I could go through the park a good portion of the way.

When I turned in at Fifty-ninth Street, the paths were lined with small grey mounds of old snow. I passed an ancient weeping willow, huge and mournful, that I could remember passing ever since I was a child. Two balloons, yellow and red, were tangled in the thin top branches of a white-striped pine. It was all beautiful and unreal . . .

I continued to walk through the park. Suddenly it was as though I could feel winter, which had seemed so strong a minute ago, beginning to end—beginning to lose its grip. It was as though in the mute brittleness of cold I could already hear the sounds of the summer that would come.

Distantly, I knew there were things I should be thinking about—decisions I should be making. But all that seemed very far away, as unreal as the stiff, unmoving figures on the sidelines when you're riding around and around high up on the tallest horse on the carousel . . .

It was just then, on the other side of the crosswalk, that I saw Mark Post and Clip West walking together, deeply involved in whatever it was they were saying. For a minute they seemed like an added aspect of my own unreality. Why should Mark Post be here now in the park talking to Clip West?

I could see West's high animation as he talked. Post stared at the ground; every once in a while he nodded, but never without keeping his eyes to the ground, never looking up. Still, every once in a while, he nodded.

I left the park quickly. How crazy it all was. To have pressed coincidence to the limits of seeing the two of them in that particular place at that particular time. Mark Post and Clip West. Nothing that could be understood; strange, inexplicable, since, after all, there wasn't any thread whatsoever linking Mark Post and Clip West. Or was there?

I took a bus the rest of the way home. It was stuffy and crowded, but I wasn't really aware of any of it. When I got home I stretched out on my bed. I told my mother I was sleepy—we could talk about Chicago tomorrow. I also told her that yes, I had had a good time. *It was this fragile thing I was carrying, and the fact that it must not be broken.*

I didn't wake up until the phone rang. It was almost nine o'clock that night. I felt that I had been dreaming, but I couldn't remember what about. I hadn't taken off any of my clothes and I realized I was hungry.

I picked up the phone, half awake, and Dave said, "They're all waiting for me in there, but I had to call you. I miss you."

"Where are you?"

"You should be here," he said, his voice fringed with excitement. "Fritz Mott's already gone through about ten cigars. He didn't really want to come, the phony cop-out."

"Where are you?"

"There's some sort of feud going on between two editors from *The Review*. One's a fag and the other one looks like Abraham Lincoln. I purposely had them seated right next to each other. There may be a murder before the night is over. I tell you, it's absolutely wild."

"Where are you? What's happening? Is it a meeting?"

"God, I wish you could see it," he said. "There are the sane ones, like Monger and Rusk, and me, of course" (I couldn't tell whether he was laughing or not), "but the publishers seem particularly demented."

"What publishers? Is Redford there?"

"Of course not," he said. "Hey, you must be kidding . . . Seriously," he said, "I wish you could be here. The moment of truth hasn't descended as yet. But it will, just as soon as I get back in there. I had to talk to you first. I'm going to let Rusk explain the whole thing to them. I want them to know exactly how important this thing is that they're doing. Christ, I wouldn't trade this minute for ten years of my life. Oh, God, how I wish you were here right now. It's a scene you'd never forget. All of these wild, crazy people in one room. The ones that are as excited as I am—who know how right and important this is—and then the ones who aren't entirely sure, who are doing it because they think in the long run it might be smarter to be on our side. And then the others—actually there are only two or three of them—like Mott, scared shitless and doing it, coming here only because they have no other choice. Oh, I tell you, Paula, you should be seeing all this."

"But what 'all this'?" I said. "*Where* are you?"

"It's a little Italian restaurant on Staten Island. Amaldo's. Nobody ever even heard of it. Except they will from now on."

"Is it a meeting? Why did you have to go all the way out there?"

He didn't answer. He said, "You can't imagine the wild day I've put in, baby. Ever since Brad called me in Chicago last night—I think I've been on the phone for ten hours. And then there were the others that had to be convinced. But it's all done now. Now it all starts."

"What? What starts now?"

"Everything," he said. "Paula—Paula, listen, I've got to go now. But I had to call you. I couldn't start without knowing you were wishing me luck. You do—don't you?"

And for just a second I was back there in that room—four-twenty-three—and the darkness at the window, and the warmth, and the touch . . .

"Of course," I said. "Now will you please tell me what it's all about? Outside of it being a matter of life and death, that is." I said it facetiously, but something instinctive gave me the cold, sudden feeling that what I was saying wasn't all that much of an exaggeration. "Of course I wish you luck," I said quietly. "But . . . Dave . . . Dave, it's impossible. I mean—" one last try for sanity—"I don't even *know* you . . . I mean anything *about* you."

"I'm six feet three," he said. "I get sore throats sometimes—but I'm a fantastic fencer. And my irresistible virtue is that I love you very much." And then, when I thought he had hung up, he said, "Oh—I almost forgot. About the pearls . . ."

"What?"

"The pearls," he said, "how long do you want them?" And then, not waiting, deciding for me, "Long—O.K.?"

"You nut," I said, "you nut." Laughing, not realizing until I had hung up that I was crying.

And afterward, alone in the bedroom in the dark, not daring to go outside where I'd have to talk to my mother, I scrunched myself up into a tight, motionless ball and thought: I love him. I love him—but I don't know what to do about it. "So, you—you better take it out of my hands. Because—if you don't—if you don't, I'm only botching it up . . ."

And it was just possible—incoherent, ungrammatical, and all—it was possible in that darkness that I might have come closer to a real prayer than I ever had before in my life.

twenty-two

THE NEXT MORNING, COMPLETELY BEwildered, I decided to treat myself to a plush breakfast at the Brasserie (I'm a great one for "comfort foods," as the expression goes). I was sitting in one of those glass-separated booths, elegantly smearing butter, marmalade and jelly onto a piece of *brioche*, when I saw Mark Post coming down the steps that lead into the restaurant. I have never seen anyone project "furtive" so unmistakably. It was played so broadly it was almost comic. Tight, tense, eyes down to the ground, he crossed to the back of the restaurant and disappeared into a corner that I couldn't see from where I was sitting.

I wondered whom he was meeting. I thought it could be Clip West, possibly continuing that very intense conversation they were having in Central Park the evening before. Then, no more than five seconds later, I looked up to see Dave coming down the very same staircase. For a moment I was sure he saw me; hardly breathing, I waited for him to come over to my table. Instead, he followed the same path Mark Post had taken to the back of the restaurant. I felt as though I were in the middle of a spy movie—but without ever having read the script. Nevertheless, I forced myself to at least pretend nonchalance as I finished my long—and expensive—breakfast.

As I left I tried, out of the corner of my eye, to find the two of them, but as far as I could see they had disappeared completely.

When I got to my office, Jack wasn't in, and neither was Pete Larsen. I got a pain in my stomach when I saw that third desk sitting there. I got up and pushed my desk back closer to the wall; then I shoved Jack's up closer to the window. The damn things were as hard to move as grand pianos, but I managed. Then I shoved Pete's desk as near to the center of the office as I could get it. Then I took down the lamp that had been sitting on one of the filing cabinets since Pete's move and arranged it on his desk and plugged it in. The whole thing was still pretty cheesy, but it might help a little so people wouldn't make any mistake as to whose office it mainly was. "The peckin' order," to use an old southern expression I had picked up from guess who.

It was just a few minutes after nine when the phone rang. It was Redford himself.

"Hi," I said, genuinely glad to hear him. "How are you? I understand you weren't feeling too well."

"Oh, I'm fine, honey," he said. "Wasn't anything too serious. How you been? You enjoy your little trip to Chicago, did you?"

There was a pause and then he said, "I've been thinking . . ." With Redford that was always a dangerous introduction. I held my breath wondering what came next.

"We haven't had a chance to chat for a while," he said. "I miss talkin' to you."

If I had been suspicious before, I was doubly so now.

"I was wondering," he continued, "why don't you slip away from that place and take some breakfast with me?"

"Now?"

"Sure," he said. "Let's see—what's a good place? I got it. How about the Delmonico Hotel? They got a nice dining room there. What you say to my meeting you there in say fifteen minutes?"

"All right," I said. "I'll be there."

He was waiting when I arrived. The dining room was all dark wood and low lights and didn't look much like breakfast. He asked me how I was but didn't pay very much attention to the answer. Then the waiter came and out of nervousness I found myself ordering another whole breakfast—scrambled eggs, bacon, the works. It was as though I could feel a problem coming and I wanted to stall as long as possible.

But it didn't work. It wasn't two minutes before he said, "Are you planning on seeing your friend, Mr. Wolffe, sometime today?"

I said, "No, I don't know—" but without giving me a chance to go on he said, "I don't know any of the exact details, but I've heard rumors of the fact that he—your friend, that is—and Brad Rusk are planning on pulling off something tricky."

I felt my breath start to come queerly.

"I've got no definite information, you understand," he said, "but there's a feeling in my bones—that and the fact of a few words I've heard dropped here and there. "So," he said, hesitating for the first time since he had started, "I wondered if you'd do me that little favor—I wonder if without exactly asking outright you could sort of find out what kind of a scheme it is that your friend's up to."

"You mean—Dave?"

"All I need is just a little information," he said. "You don't have to get any exact details." He was quiet then, looking at me directly, and I knew I should say something, but I didn't know where to begin. Finally he said, "The reason I haven't been in the office is that Bob feels that the plans your lovely *friends* have include discrediting him by tearing down his team, one at a time."

"Aren't you coming in today?" I asked him, compulsively gobbling away at the food in front of me.

"Bob says it'd be better if I said I wasn't feeling too well for a couple of days . . ."

Then he squared those elegant southern shoulders and, out of what seemed like sheer braggadocio, he said, "For my part I couldn't care less. It's just that I don't want to stand still for their killing Bob like that."

He drank some of the coffee, the only thing he had ordered

for himself, and, seeing the cup shake in his hands, I wanted to look away, but I forced myself not to.

He said, "It's not as though it's a matter of life and death with me the way it is with some people. Look, the way I figure, whatever goodies there was to be tasted, I done my fair share of the tasting already—and more to boot. Who else you know waltzes himself right out of a Mississippi shack into the Colony?"

He looked at me as though half expecting that I would contradict him. "Any day all of them goodies come to a stop, it don't make no difference to me. 'Cause I already had it all. Anybody interested, they better remember that."

It was almost more than I could take, but again I forced myself not to look away. It was about time I stopped running from things, playing hide-and-seek with life, and if I had to begin by watching one man's coffee slopping out of a cup, at least it was a place to start.

"You'll find out then?" he said, trying to keep the desperation from showing in his voice.

I felt my throat go dry as dust; when I spoke I could only dimly hear the words muffled in my head.

"Well, I—I don't think I could do that."

"You ain't bein' serious?"

"Yes." The muffling in my head was even thicker this time; I had to struggle to make some spit.

"Oh . . ." he said. It was like watching the air go out of a red balloon.

I didn't know what to say, only that I had to say something. I kept fumbling around for the beginnings of sentences, not able to find one that I could grasp onto. Finally in desperation I said, "Well, if you feel that's their plan—to tear down Bob Moshier by discrediting the people that he's hired, why don't you do something about it?"

"How do you mean?" he said.

I didn't have the slightest idea what I meant; all I knew was that I had to keep talking.

"Well," I said, "why don't you fight it by proving what you *have* done? By proving how good Moshier's people are? They can't discredit him if there's nothing to discredit. Why don't you just give him enough ammunition to show him how effective the team he hired *is?* Maybe even tell him what you believe is wrong about the way Rusk operates."

It was the first time I had ever seen Tom Redford at a loss for words. He didn't even say, "How?" He just sat there with his mouth open, waiting for me to go on.

So I went on, making it up as I went . . . "If you just wrote a paper," I said, "a sort of summary of all that you've done as publisher since you arrived, plus your plans for the future. What I mean is," I said, continuing compulsively, "show them exactly all the things you've done for *Today*, all the steps you've taken to make it better since you arrived——" And all the while in the back of my head, a little voice kept saying *Like what—like what?* And I could see Redford's eyes beginning to come alive again.

"Hey," he said, "I just got me an idea . . ." He beamed. "Why don't *you* go to Bob?"

I said, "I'm sorry?"

"Why don't you let me call Bob and see if he thinks it's a good idea—and then you could go right to him and tell him what you've just told me . . ."

"You mean *me*—go to Bob Moshier?"

"Sure," he said, nodding his head to show how sure he was, smiling, beaming, but the desperation nevertheless showing itself, not only in the suggestion but in the tone of his voice.

"If you went—providing Bob agrees, of course—you could do just that; you could speak to him in regard to what you—as a staff member—feel I've contributed to the magazine. It'd be perfect."

"You wait right here," he said, his persuasive con-man best. "Don't go away now." Getting up, smiling, charming, leaving me without anything to say, or think, except the awareness of how once again I was doing just great at acting as my own executioner. Because, after all, where had he gotten the loony idea in the first place? From me.

Sitting at that table, waiting for Redford to come back from his absolutely insane phone call, I wondered who I was and where I was coming from, and why I was evidently in the heart of this thing where I had no place being—where, furthermore, I not only didn't belong, but where the breed was one completely foreign to me, different, except I had this compulsion to keep pushing my way into the middle of them—and for what end I didn't know—I had never known.

And suddenly the significance of Dave's call last night struck me full force. Whatever had been going on, there was no doubt in my mind now as to how important it had been. I tried to figure out what it might have been. Something had happened. Something very important. Something enough to scare Tom Redford so much that it showed.

At that moment Redford returned. He sat down, and I could tell immediately that Bob's response hadn't been what he wanted. Nevertheless he tried to carry it off.

"I'm sorry I took so long," he said. "Bob was in a meeting."

"Did you get to talk to him?"

"Yes," he said, falsely hearty, "sure, sure I got to talk to him."

And then, still trying to keep it light even though the world was Chicken-Little falling around his head, "He thought maybe it wasn't too good an idea for you to come to his office right this minute." And, suddenly, almost as though I had been there, I had the full realization of how Moshier had reacted to the whole suggestion . . . Redford's telling him I was a "friend" of Dave Wolffe's, suggesting that I go there right now, tell him all the wonderful things at least one member of his staff knew he was doing as publisher—and all the while, the fantastic lovely little touch that must have been going through Moshier's mind, the fact that his office was directly next to Dave's.

"Well—I guess I better get back," I said, realizing again that I hadn't even told anyone where I was going.

But it wasn't over.

"Are you going to find out for me?" I think that what threw me most was that there wasn't even a trace of an accent. This wasn't Southern Tom anymore. This was something else. Something quiet and simple, and very, very lethal.

"I don't think I could do that," I said.

"What do you mean?"

"I don't think I could—what I mean is . . . it would be like spying."

"How come you never thought of that before?"

"What do you mean?"

"Where you went to dinner . . ." he recited, mockingly sing-song, "and the violins, and the limousines, and what he said, and practically the number of times he went to the john. How come none of that ever bothered you before?"

I didn't have anything to say. Because I didn't know. Just because. Because what he said was true. But because it was different now.

"What's the matter, honey," he said, "you all of a sudden afraid of guilt by association?" The words were as cold and tearing as a sliver of tin.

"What's the matter?" he said. "You're a pretty tough little Wop kid, aren't you? I mean, underneath all that outside sweetness and being good—you don't fool me for one second. You're really a shark, a killer. Only I'm too tough for you, baby. I could kill you before you even knew what happened." We looked at each other in absolute silence. Finally he said, "What's the matter? You just going to sit there and stare at me for the rest of the morning? Well? . . . are you?"

"I was just wondering," I said.

"Wondering what?"

"Whether you had any more knives on you."

It was hard to pin down the name for the laugh that exploded in that dark-wood, white-tableclothed dining room. It was hysterical—desperate; it was a laugh that could kill you.

"Look, missy," he said, "you look at me now and you remember what I say. I ain't no loser. Never in my life. I'm a winner. If I hadn't been a winner I'd of been killed at least forty times in my life before. You got any idea what my bringing up was like? I'm a genius at survival, honey—didn't you know that? And if I ever did see any chance of my being killed in this situation, I'd bring the whole goddamn crumbling world down with me. You, Bob, the hell who . . . Now you better remember that."

I shivered and didn't say a word. And then, two seconds after the last knife had been thrown, Redford was all charm and sweet-faced smiling again.

"Have one of those Frenchy-type rolls, why don't you?" he said, all graciousness wiping out the absolute threat of what had been said.

"No, I guess I better get back."

I watched him as he paid the check, smiling at the waiter, and then he came over quickly to help me with my chair, and we were out of the restaurant, standing on the cold sidewalk where the first drops of a grey, biting rain had started to fall, when he said to me, "Why don't you call your friend as soon as you get back—no reason to waste any time." And while I stood amazed, completely disbelieving, he went on, "Tell him you have to see him. Tell him you have a problem."

"But—I told you—I can't."

"You're a woman," he said, smiling, not even listening. "You'll think of something." And, you could see it—he really believed it was all that simple.

"Well," he said, "I think I'll drift over to Saks. I'll walk me around some of them faggy-fancy stores for a while. I'll talk to you," he said.

"But I can't call him."

And he said, abruptly, quietly, "I'll call you. This afternoon . . ." And then, very quickly, he walked away from me, heading toward Saks—toward the "faggy-fancy stores" where he'd look around, elegant, aloof-looking, maybe even joking with one of the salesgirls—the picture of a "winner."

I turned after leaving him, and he waved at me, casual, easygoing, but just for a second I saw the mask slip, saw a man maybe without a job, told to stay away, never completely able to wash out the image, the spur of his beginnings. A man in the rain with no place to go.

I went back to the office. The rain was coming down pretty hard by the time I got there. Pete Larsen, my secretary told me, wasn't coming in today. In the meantime Clip West was losing no time. He was all over the place at once, chatting with the salesmen, calling in writers, studios, dictating memos . . . but never looking you in the face when he talked to you—when he spoke about his friend, his neighbor, Ted Monger—"Monk."

And I sat, doing whatever it was I did mechanically, not really there, and went out to lunch the same way, like a toy with a key in its back, going through the motions, never able to think one minute past the minute it was. Ten o'clock . . . eleven . . . And then one . . . and two . . .

The phone rang. Redford said, "Did you call him? Have you seen him?"

"No—no, I haven't."

"Why? Isn't he in his office? When did they say he was coming in?"

"No . . . no, I didn't call him."

"When then? What are you waiting for?"

I didn't say anything.

"This is your last chance," Redford said. And I didn't know what that meant. Except I was more frightened than I could remember having been, and I wanted to lie, I wanted to say, yes, that I had tried to find out, that I hadn't been able to reach him, but that I had tried. And instead I said, "No. I can't ask him."

"And that's your final word?" he said, the words like shiny, quick-tipped knives.

"Yes," I said. "I'm sorry but—I can't ask him."

"Then go to hell," he said, "and don't forget. Don't forget I warned you."

And then he was absolutely silent, and I remembered Jim Canfield—"*You better know who you are. Just pray you end up knowing who you are. God help you if you ever forget that . . .*"

And I said, "I'm sorry. But I meant it. I'm not going to try to find out," and then there was nothing else, not even an argument, just silence.

I wondered what would happen next. Frightened about it. And whatever it was would be related to the strange phone call from a restaurant on Staten Island—and editors gathered there—and publishers and, strangest of all, "*I needed to know—I needed to know you were wishing me luck.*"

I thought, this is the way it must feel when an earthquake is beginning. The first rumblings, the dry taste of fear, the disbelief, the desperate need to think it was all simple, that it would pass, that it was only in your imagination, that it would all disappear. But knowing better, knowing it was just a matter of time—of minutes maybe. Of minutes at the most.

Or—as it was for me—the very next morning.

twenty-three

"HERE'S SOMEBODY YOU KNOW . . ."

At first, half asleep, groggy from dreams filled with threatening people, with terrible words—none of which I could any longer remember—I looked at the face of the little old man at the newsstand in the Kimberly lobby from whom I had started to buy my paper every morning. For the first time I seemed to notice how round his tiny pink face was; I saw the dark smudges that years had imprinted under his eyes.

He said, "See," holding out the paper for my inspection.

The caption read, "Kimberly Company president accused of mismanagement by top magazine executives." And underneath the picture it said, "Robert Moshier, president of the Kimberly Publishing Company, was late last night brought before members of the Board of Directors. He was accused of gross mismanagement of the company's affairs. Formal presentation of the charge was made by David R. Wolffe, head of the Magazine Division, and Bradford Rusk, policy coordinator. . . ."

"See," the little man said, "I knew you would know them. There's a lot of mishigas going on in the company, isn't that right?"

And I said, "Yes—yes, I guess there is," and put down my fifteen cents and picked up the paper, walking like someone drugged, over to the elevator.

When I got to my office I sat down calmly, first taking off my gloves and putting them into the drawer and setting my purse to the side, and then finally sitting down and opening the newspaper as I had done a thousand times before, and beginning to take in what was there, with all the pretended calmness of reading any other front-page story.

The account read: "In an almost unprecedented incident in the history of magazine publishing, two of the top executives of the Kimberly Publishing Company brought charges of mismanagement against the company's 42-year-old president, Robert Moshier. Introduced at an emergency session of the company's Board of Directors, the formal charge, made up of fourteen points of grievance, was presented by David R. Wolffe, head of the Magazine Division, and Bradford L. Rusk, policy coordinator for the entire company.

"The report, which was not made public in full, stated that on fourteen counts Moshier was found not to have acted in the best interests of the company of which he is president. Both Wolffe and Rusk, at a special meeting of the Board, requested by Wolffe, asked for the immediate dismissal of Moshier.

"The event—a startling one in the history of the 70-year-old traditionally staid publishing company, was made more dramatic by the manner in which the request for Moshier's dismissal was presented. Taking the form of a twenty-page indictment, the document was signed by twelve top officials of the company, including the editors-in-chief and publishers of each magazine. The single exception was Thomas Redford, publisher of Today.

"An especially bizarre touch was added to the incident by the rumor, unconfirmed, that the signatures were collected at a secret meeting called late on Tuesday night at a small Staten Island restaurant."

I sat down and for five full minutes just stared at the wall in front of me. It wasn't that I was thinking anything. It was just five minutes of complete, incomprehensible, numb staring. And from that point the whole unreality of the day took shape, incident after insane incident.

The first thing that happened was Jack Sheehan calling. He had a cold and he wasn't coming in and would I please tell Pete Larsen. Five minutes later, the phone rang again, and it was Pete Larsen. He had broken a tooth and he wasn't coming in, and would I tell Tom Redford. Neither of them mentioned anything about what had been in the paper that morning.

I went out of my office and headed for Jane Perry's to tell her about Pete's not coming in. Halfway down the hallway three secretaries were gathered in a tight, buzzing cluster. One of them had a copy of the *Times* with Bob Moshier's picture on the front page. The murmuring stopped as I approached and then started up again right after I had passed them. That was the way it would be all day. No matter where you went, the air would be pockmarked with gossip and conjecture and apprehension and speculation. And underneath it all—from the file clerks to secretaries to executives—there would be the same basic fear: What if I lose my job? It didn't matter whether they made $65.70 a week take-home pay or whether the position dragged in thirty thousand a year, plus membership at Wingfoot and The Sky Club; the basic terror was the same. Make waves in the company, especially a large company, and no matter how big the individuals are that are making the waves, every employee immediately thinks, What about *me?*

It's a special kind of fear that spreads like wildfire. And it doesn't necessarily have to start out of any *real* happening. Just a word can do it, a rumor, a comment dropped in the elevator. And in this case there was more than that. Much more. It had all the ingredients of the best late-late gothic suspense movie they had ever gotten goose-pimples over. And, best of all, any blood that was shed along the way would be the honest-to-goodness red-and-white corpuscle stuff, and no cinematic imitations, no matter how realistically approximated.

There was just one hitch. Part of that blood—that real blood—might be their own.

I said to Jane Perry, "Pete Larsen just called. I'm supposed to tell Tom Redford that he won't be in today."

"Fine," she said, smiling with mischievous coolness, "if Mr. Redford calls I'll tell him."

"You mean you don't think he'll be in even today?"

"Not if he's still obeying orders from his buddy, Bob Moshier. As far as I can tell, our southern friend had strict instructions not to put a foot within a block of this building."

"Even after *that?*" I said, motioning to the paper open on her desk.

"As far as I know the instructions were no matter *what* . . . What do you suppose is going on in this nutty company anyway?"

"That's what I expected you to be able to tell *me*."

She said, "Do you want to talk to Mr. Redford if he *does* call in this morning?"

I hesitated for a moment. "No," I said. "Thanks, no. I don't really have anything to tell him."

"Paula . . ." And when I stopped she just looked at me, as though weighing whether she knew me well enough against what she was going to say. Finally she said, "I have a problem. And I don't know what to do about it."

"What is it?"

She hesitated for a moment and then said very quietly, "It's about Mr. West . . ."

"Old buddy Clip . . ."

"Well," she said, "it's really none of my business, but, well I do work for Mr. Redford—and I guess I'm even fond of him—in a kooky sort of way. The thing is—I don't think West realizes how thin the walls are in this place. What I mean is, he's right there in the office next to mine, and I don't think he realizes that you can hear every damn word he says on that phone—whether you're interested in listening to it or not."

I laughed. "Heard something juicy, have you?"

She said, "This is worse. It's about Mr. Redford. I can hear him talking about Mr. Redford."

"I guess he ought to," I said. "After all, it was old buddy Tom that got him the job. It must have taken a little wire-pulling to do that."

"You don't know the half of it," she said. "I shouldn't be telling anybody this—but do you know that Clip West is being paid more money than any promotion director in this whole town? You have no idea how many memos went out of here to David Wolffe justifying the salary that Redford was asking for his friend West. He never even sold *himself* as all-out as he sold Clip West."

"I guess that's what they mean by southern solidarity . . ."

"That's just it," she said. "That's why the phone calls throw me completely. The first time I thought it was all my imagination—or a joke—or something like that. It had to be. After all, Redford had gotten him the job, and he hadn't even been here more than a day. But it keeps happening."

"What keeps happening? . . ."

"These phone calls between Clip West and Ted Monger. Did you know they knew each other way back when?"

"Yes, I did," I said. "I found out that day he took Pete and Jack and me to lunch."

"Do you know what West says about Redford, his buddy, the guy who got him the job? He says how stupid Redford is, how he'll never recognize what's happening until it's all over . . . at least seven times a day. It's like they practically keep an open wire between them." She stopped. "So that's what I mean—I don't know what to do."

"You haven't heard from him at all this morning? I mean not even after what happened last night?"

"No," she said, "and in a way I guess it's just as well."

"Isn't *anybody* coming in today? First Pete and then Jack—and Redford . . ."

"Our friend Mr. West is in," she said. "Bright-eyed and bushy-tailed and out to make friends as soon as possible." Just then her phone rang. We both held our breaths to see if it was Redford, but it wasn't, and as she finished taking the call I left her office and started back to my own.

She was right. When I went back past Clip West's office, he had four of *Today*'s space salesmen in there—and he was making points like crazy. I heard him say, "You gentlemen all such *pro*-fessionals—you got a right to have some real first-class *pro*-fessional promotion backing you up . . ."

I was too restless to sit at my desk. I roamed around the hallways on nonexistent errands, I walked in and out of stockrooms looking at the supplies; finally, I took the elevator and just sort of strolled from one magazine area to another. As casually as I could, I checked to see whether any of the people who had signed the petition were in their offices, but the place was such a mess that it was hard to find out.

Outside of Ted Monger's office I just stopped and blatantly asked his secretary whether he was expected in. She said she didn't have any specific information and she didn't know what Mr. Monger's plans for the day might be. Miss Ice Pack herself.

No matter where you turned, you walked into little knots of people whispering about what had happened. Here and there someone would be trying to concentrate on whatever he was doing. I passed the editorial offices of *Today* and saw Charlie Ward, one of the old-standby editors who had been there way back in the beginning when Canfield had started the whole thing, an irresistible little pixie of a man, an editor to his fingernails, with wonderful pink cheeks and a habit of passing along the latest of whatever dirty stories he had heard, and then blushing deeply as he delivered the punchline. Now he was deep at work at his desk and I stopped in the doorway and said, "Hi."

"Oh, hi, honey," he said, "what do you think of the little fun and games we got going here? Just one big playground, right?"

I said, "It's wild, isn't it? What do you think's going to happen?"

"Search me," he said, and then, with a delightful pseudo-leer, "Better yet, let me search you." He got flame-red as he said it.

"You're the only one that seems to be working around here," I said.

"Got some proofs to creck," he said. "The goddamn magazine closes in two days. Wouldn't you think they'd have sense enough to put off their games till sometime when we weren't so busy?"

I laughed at him and said, "Guess they're not as conscientious as you are."

"Conscientious my foot," he said. "They just got a power bug up their ass, that's all. Fool amateurs—why don't they get themselves an erector set if they're that hot on kids' games . . . Or go get laid every lunch hour—that'd settle them down quite a bit."

"I'll let you get back to your work," I said.

"Listen, *tressora*," he said, "for you I've not only got the time but the money as well . . ." At which point he blushed so deeply even his ears turned red.

I threw him a kiss and got out of the room so he could get back to his page-proofing.

I kept on walking around, as though I had some aim in mind. The company was like a series of little vignettes in an experimental Italian movie. Funny, and tragic, and wild, and beautiful, and mostly pretty improbable.

I ran across Bob Warnick, who was head of special projects for Kimberly, whatever they were. He had a copy of the newspaper under his arm, and when he saw me in the hallway he started on a conversation, just as though we'd been talking all along. "Goddamn stock market," he said, "haven't been able to get through to my broker all morning. Probably gone to Cuba or someplace . . ."

"You interested in Kimberly stock, Mr. Warnick?" I asked, and he said, "Listen, I got two houses in Westchester, a Piper Cub and two filthy kids going to Northwestern. *That's* how interested in Kimberly stocks I am."

The day kept exploding in little incidents like that—little Fourth of July firecrackers that you react to and enjoy, but all the while knowing that it's going to be more than this, waiting for the big baby to explode, getting tenser and more stiff-mouthed as you wait.

When I passed the editorial area of *The Review*, most of that magazine's editors were gathered in the conference room. At the front of the room there stood a colossal red-headed man in a white shirt with the sleeves rolled up above the elbows. He was absolutely immense, and he looked as if dozens of red Brillo pads had been tied to his skull. He also had a beard. Immense and fiery red.

I wondered who he was, but when he saw me standing in the hallway he walked directly back to the end of the room and closed the door. It was done so easily and so authoritatively that I didn't even react until I found myself facing the blank closed door. Whoever he was, you had to hand it to him; he certainly kept his cool. Considering his surroundings, that was quite a feat.

Right after lunch I went to Jane Perry's office to see whether Redford had phoned that morning. He hadn't.

I managed to make myself sit at my desk for an hour, and then I took off again on my wandering tour of the place. As far as I could see, none of the people who had signed the petition were in the office. I passed the editorial area of *The Review* once again, and this time everybody was scurrying around like bewildered little rabbits. The Women's Editor, a very grand, white-skinned type, was on the phone to someone saying, "But, darhling, I tell you I must have those gowns for photographing tonight—I know you haven't planned to give anyone even the teeniest peek until next Monday, but I tell you, Rudolph, this is an emergency . . . We have this absolutely *freakish* new editor . . ."

The Fiction Editor, a tiny elf of a man who always looked drunk, even the first thing in the morning, was walking up and down the corridor with a glass in his hand that was either pure Coke or straight bourbon saying, "But *how* can I get a ten-thousand-word story from Gore Vidal by tomorrow morning—and it has to be a love story about a hare-lipped girl and a club-foot boy and it all has to take place in Nome, Alaska—I mean, I don't think Gore Vidal ever wrote a *love* story, did he?"

Back at *Today*, Clip West was still going through his wooing of the sales staff. "Why don't we all go have a drink together after work, O.K.?" he was saying to the man who was in charge of all food advertising. "I got a few promotion ideas I'd like to get your very valued and expert opinion on."

I wasn't able to sit still for very long at a time. Near the end of the afternoon I drifted up to were the teletype machine was kept. It had always fascinated me, that squat little monster of a machine, quiet one minute and then suddenly, eerily, spitting out little snips of words and sentences. As I watched there was a release on a bus strike which had just been announced in Boston, a race riot in San Diego, a whole American battalion that had evidently been wiped out in Saung Kow, as yet unconfirmed . . .

It always made me jumpy watching that machine. I think in my subconscious I really believed that the machine wasn't just announcing all these disasters but *creating* them.

Finally, in desperation, I went down to the lobby to pick up some cigarettes. As I came up to the counter the little man behind it gave out a great whoop of laughter. "*Bu*-ti-fool! It's *bu*-ti-fool!" He turned to me as though eager to share the news of this glorious "bu-ti-fool!" thing that had happened, and he laughed again and said, "Oh, it's you," and then, to another little man—the Fiction Editor—standing at the counter, "Tell her, tell *her*, let her see how bu-ti-fool it is . . ."

"Well, it's like thish—this," he said. "First thing this morning all the editors on *The Review* receive messages. There's to be an important meeting in the conference room at ten-thirty, and be there. Well, I don't have to tell you what the atmosphere around the office was like this morning, do I? As cheery as an open grave, right?" He looked me straight in the eye—or as closely as he could manage. "So, comes ten-thirty we're all in the conference room—everybody looking 'pale under their tans' as the expression goes. Not that we're not all 'secure,'" he said, "it's just that some of us are a little less secure than others. Anyway, promptly at ten-thirty this wide mountain of a man comes into the conference room, a huge sheaf of papers in his hand. He goes right to the front of the room and he says, 'My name is Pete O'Mara. Since you're the staff here I wanted you to know before word got out publicly. By tomorrow morning it will be officially announced that the Board of the Kimberly Publishing Company has unanimously voted that I become the new editor of *The Review*—effective immediately.' Well, there were a few sort of strangled gasps in the room, but after what had been announced in the newspaper everybody'd read that morning, we wouldn't have been too surprised to hear that Spiro Agnew had become beauty editor.

"Anyway," the little man went on, "this red-headed giant didn't waste any time on settling in. He said, 'I understand we have two and a half days to close the April issue—so in the cause of efficiency and expediency I've allotted a half hour for each of you to completely familiarize me with every story appearing in the issue. I want to see full copy, captions and all photography. As I said, each of you will have exactly thirty minutes to explain, illustrate, and, if necessary, justify those pieces for which he is responsible. A checklist of the sequence in which I will see each of you individually is at this moment being posted outside my door.' Then he kind of ran his hand through the mass of prickly red hair and said, 'I'll expect the first one of you in five minutes,' picked up the huge, messy-looking sheaf of papers he had come with, and started to the door. Then, when he was almost there, he stopped for just a moment. 'By the way,' he said, 'if you were thinking of eating lunch out, I suggest you start thinking otherwise. Oh—and it might not be a bad idea to tell your wives not to expect you till they see you. Eh, those that have wives that care, that is.' At which point he very grandly left the conference room."

The newsstand man was leaning on the glass counter, transfixed by every word. "It's like this all the time—the magazine business?" His eyes opened even wider in the delicious prospect of still further insanity to come. "Wait . . . wait," he said to me, "wait you should hear the rest."

The little man who was telling the story took a deep breath and went on. "Well, to put it in succinct terms, it was like slaughter time in the sausage factory. One by one each of us went in to tell him about the features that were our personal responsibility, to bring him the copy, to show him the photography or art work that was being used to illustrate the pieces. In each case he killed whatever had been gotten ready for the issue —a child-guidance feature by Doctor Spock—forget it, he wanted a ten-thousand-word piece by Jackie Onassis by tomorrow morning, a piece on how to raise step-children, mind you.

"The food feature had been on tailgate picnics. He wanted secret exclusive recipes from ten of the topmost chefs in the country, and all to be photographed in full color by Andy Warhol.

"The fashion pieces he killed completely—instead he got right on the phone directly to Cardin in Paris—two full spreads on his latest collection, and then, on to Production—directions that the latest beauty pages were to be printed on silver lamé stock. For six hours he sat in Brad Rusk's old office, his feet up on the desk, his mop of red hair gleaming like a devil's halo—and he pulled the entire issue apart. He pulled out stories completely, he commissioned free-lance photographers, he told two editors that their salary was going to be doubled by the end of the week, and two other editors he fired. And all just like that." He snapped his fingers crisply.

"And now we come to the best part," the Fiction Editor said. "It wasn't until four-thirty this afternoon—and then only by the slimmest of chances did the beautiful performance tragically end."

"Performance?"

"It seemed he tried to get a five-hundred-dollar advance from the cashier—he sent his little secretary down to the cashier with a voucher. But cashiers, as you know, are not like any other people. Cashiers do not believe the day is Sunday until it has been checked, verified and put through a computer. If it had not been for the cashier the beautiful impersonation might have gone on for—well, who knows how long?"

"You mean—you mean he wasn't really the editor?" My open mouth at this time matched that of the awe-stricken little counter man.

"A beautiful performance," the elfish Fiction Editor said, "full of, as they say, *panache*. Not just beautiful—glorious . . . A glorious performance. And if it hadn't been for greed—for man's ever-present cupidity, who knows how far it might not have gone."

"Then—he wasn't the editor at all?"

"He was, as a matter of record, a very hard-nosed newspaper columnist who just happens to hate Brad Rusk's guts. And what better way to display that hatred, what more beautiful ploy than to come in on the one day when nobody knew who was who or who had been fired or hired or had died or what-not—when the whole climate had been perfectly prepared for him. Walk in and take over—I've heard that phrase all my life but I never knew how true it was before."

"With his feet on the editor's desk—" The newsstand man exploded into fresh and delicious laughter, "and he gives orders to everybody—do—and do—and *do*—and everybody does! Oh, it is bu-ti-ful."

The two of them were fully enjoying their hysterics after I bought my pack of cigarettes and started back upstairs.

The atmosphere of uncertain chaos was the same at it had been since this morning, except worse now, because more time had passed.

I decided to get rid of the rest of the time that was left by going into the ladies' room and getting ready to leave. But I had no sooner started combing my hair than two secretaries came in together, each carrying her own little can of Spray-Net.

"But what do you *really* think is going to happen, Joan?" the blonde, cooky-mouthed one said.

And Joan said, "How do I know? Besides, maybe we'll all have new jobs or something by next week."

"What makes you say that?" the other one asked.

"Oh, I been in places that folded before," Joan said. "This Feldman Hat Company that I use to work for, you should of seen what it was like just before the whole thing folded. Something just like it is around here now—everybody jumpy and worrying, and rumors going on about everything. I swear, it smells exactly the same way as it did at the Feldman Hat Company."

That was about as much as I could take, so even though it was a little early, I decided to go home. I had it in the back of my mind to tell Clip West I was leaving a little early, but when I went past his doorway he was deep in a telephone conversation with somebody, his chair swiveled around so that he was facing the window, and his shoulders hunched over the telephone. I decided the hell with it, and went.

The subway was crowded as usual, but I managed to wedge myself behind one of the poles where at least I could have something to hold onto. For a while I read the ads around the top of the car.

When I had exhausted the subway literature, I turned sideways and started to read a copy of *The New York Post* that the man next to me was holding. For about five minutes he was deeply engrossed in a very vivid account of a love-nest killing in Jackson Heights, so I read it along with him, even though I didn't especially enjoy some of the gorier passages. After that he turned to the want ads at the back of the paper. But after a minute or two he evidently got bored with them and went back to the front of the paper. We had just gotten into the Fifty-ninth Street station, and there was the usual crowd of people waiting to get in, so that it took him a minute or two to get the paper under control. A woman with a purple spangled hat and four Bloomingdale's shopping bags had managed to wedge herself between me and the man, so that for a moment the paper was hidden from me. By the time I could see it again, the man had it opened to the second page. For a wild second I caught a glimpse of two faces, and my breath stopped—and then the train lunged and I couldn't see them anymore. Completely ignoring the mutterings and stares of the purple-hatted woman, I pushed closer to the man holding the paper. Under the spread of the three faces there was a boldfaced line that read: Three Top Executives Dismissed From Kimberly Publishing Company. The pictures were of Bob Moshier, Brad Rusk and Dave Wolffe.

Unaware of doing so, I pushed closer to the man. The story started: "In what was without a doubt the greatest coup ever attempted in the history of magazine publishing, its two leaders, Brad Rusk, policy coordinator, and David Wolffe, chairman of the Magazine Division, found themselves the victims of their own plot. Ironically enough, the man they tried to oust, Robert Moshier, super-salesman and president of Kimberly, was also suspended from the company.

"The whole uprising started on Tuesday night, when Rusk and Wolffe, calling a secret meeting of the top executives of all of Kimberly's publications, managed to get the signatures of each of these executives on a petition which would present to the Board of Directors their reasons for demanding the dismissal of Moshier. The meeting was held at a small suburban Staten Island restaurant. Present at the meeting were—" With one flip motion, the man folded the paper and stuck it under his arm. We were coming into the Seventy-seventh Street station, and he was making his way to the door to get out.

For a second I had a wild impulse to ask him if I could have the paper, but the train had already stopped, and the man was gone.

The rest of the way home I tried to figure out whether I had imagined it all. And by the time I got to my station I had almost convinced myself that I had. Except for one thing. The look on Dave's face. I had seen that look before. But only on the faces of children. It was a look of absolute hurt and pain, and betrayal. A look that wiped away any possibility of future trust in anything—or anybody.

I bought a paper as soon as I got off the train but—now that I had it—I couldn't make myself open to that page again. It was as though all I could see was Dave's face, blunt with the look of absolute cold-eyed bitterness and hurt.

As soon as I got home I tried to reach him—at the office, then in Chicago, then at a Forest Hills number that I found in the telephone directory—a number where a woman's voice said, "No—no, I'm afraid David isn't home. He hasn't been here all week. Who is this? Can I have him call you when he does come?" I hung up without answering.

I remembered the speech. It seemed very old and a long time ago. "*Aren't you afraid you might fall in love with me? It could happen, you know . . .*"

I went back to the paper and made myself open it, all the time avoiding the sight of the bitter, hurt eyes, the mouth gone cruel. I read straight through the write-up. There wasn't anything in it I didn't already know. Next to the write-up there was a small editorial box. It said, "Who knows what the next step in this publishing debacle will be? After 70 years of staid business procedures, it was completely unlikely that a single day at the Kimberly Publishing Company would include a concerted rebellion of most of its top editorial and business people—an ultimatum demanding the ouster of its President—and, within the same day, the dismissal of all three of its top executives. Unlikely—except that it happened. And—as for tomorrow—who knows? As they were given to saying in some of those great old-fashioned cliff-hangers that used to run in The Review: watch for the next exciting, unexpected installment. We will be watching, gentlemen."

Compulsively—my hands shaking—I dialed Dave's office again. When nobody answered after it had rung seventeen times, I hung up.

twenty-four

THE ATMOSPHERE IN THE COMPANY THE next morning was like unadulterated Good Friday. People sort of groped their way around, and if you came up behind them, they gasped and swung around at you.

Both Jack and Pete were back, neither of them saying anything or even really looking at you. I asked Pete how his tooth was and he stared at me in bewilderment and said, "Tooth? Oh, oh, it's fine, Paula." And that was the first time I realized that he was probably looking for a job somewhere else. Evidently even calm little rabbity Pete Larsen could be pushed just so far. If Redford had just fired him when he decided to make his friend Clip West the promotion director, it would have been simple. Too simple, maybe. Instead they shoved him into an office with two other people and impressed on him how great he was for remembering to provide pencils and Life Savers and gum at sales meetings. All so that, eventually, they wouldn't have to say they had actually fired him. It was probably what Redford thought of as being "humane." All it really meant was that Pete would lose out on any severance pay he might have coming, which was probably considerable since he had been there so long.

As for Jack, from the looks of him he had obviously been on a drunk the day before. The skin on his face was white and pulled taut across his bones, and his hands trembled.

He said, "Do you suppose this is what Mr. Wolffe had in mind that night in Chicago when he said he wasn't putting any of his own money into Kimberly stock as yet?" His laugh had a kind of empty, hollow sound to it, but it was good to hear nevertheless. God knows, there didn't seem to be anybody else laughing around the place. With one exception that is—Clip West. Mr. West marched up and down the hallway, stopping every *Today* salesman he could find and delivering his shiny little southern-embroidered speech that went, "Hey, fella. I hear from around that you're a real crackerjack salesman. No kidding, I heard it at least three times yesterday. And all from top people. I got it in mind to do a special promotion just for you that's gonna knock them on their asses. Hey, buddy, what do you think of that?"

After the third time he had delivered the speech I was absolutely fascinated. I found myself repeating it with him word for word, and, sure enough, it didn't vary by a single syllable. I'd hear him make the speech out in the hallway, and then, after it was over, I'd watch the different salesmen as they went past my office. They'd be grinning from ear to ear, each and every one of them, just as though Jesus Christ had come down and given them a raise. I remembered what Jane Perry had said about West knifing Tom Redford and, just for a second, it didn't seem as completely far-fetched as it had the day before.

I passed the office manager, Helen Walsh, in the hallway. She was laughing a little hysterically, and when she saw me she said, "Would you believe it—they're fighting over the *furniture?*"

"The furniture?"

"Rusk's and Wolffe's and Moshier's," she said. "One of the editors wants Rusk's teak desk and somebody else is trying to bribe me to get Wolffe's brocade couch. Would you believe it? I finally had to take my phone off the hook just so I could get some sleep."

"You mean they called you at home?"

"Honey, like every five minutes," she said. "And then this morning it started all over again here—like a quarter of eight. I didn't even get a chance to get to the damn john."

But most of all, the entire morning, in the middle of all the weird, heart-breaking, frightening, funny things that were going on around me, I couldn't bury the aching need to find Dave.

Every time I did something to try to reach him, and failed, I got more and more desperate. He wasn't in the building. And he wasn't expected. I found out via Helen Walsh—playing it as phony cool as I could—that a shipping service was coming to take away all his personal things. After that there wasn't anyplace else left to call. Except his home. And I didn't think I could go through that again. Every time I had to go to the john, I'd hurry back to ask my secretary if there had been any calls—and every time she said no I'd find myself saying something stupid, like, Are you sure?

So that finally, all the doors tried and all the doors found locked, I found myself with a drowning, desolate pain of loss. "*What if you fall in love with me, and then after a while, without ever saying why, I might just decide not to see you again, and you'd start trying to find out why and you'd keep calling me and leaving messages, only I'd never answer any of your calls* . . ." He had been laughing when he said that, but now—

I tried but I couldn't stay at my desk. I drifted back to Redford's office, but Jane Perry said there hadn't been any word from him. I walked all around the building. On the bulletin board outside the editorial offices of *The Review*, someone had posted a sign that said: *BRAD RUSK SAVES.* The movers were carting things out of Dave's office. One of them said, "Wasn't it just a couple of weeks ago we dragged all of this stuff in here?" and the other one said, "Easy come, easy go."

But *why* would he want not to speak to me? What reason was there? Even in the middle of the thing in Staten Island he had called me. "*I had to know you wish me luck. You do, don't you?*"

More and more the whole thing felt like being in the middle of a building that has started to crumble. The mailroom boy was saying to Rusk's secretary, "But what do I do with all this mail? You got a forwarding address?" And she said, "Why don't you just throw it away, Otto; he's not with us anymore."

It was almost lunchtime now, and the building was taking on its usual twelve-to-two abandoned look. I wondered if I should send out for a sandwich or something. I couldn't stand the idea of going out and eating by myself. I kept walking around and around the emptying offices feeling like the last rat on a sinking ship.

Still roaming without any particular direction, I was so startled by an unexpected sound that I could feel the muscles in my stomach react. My heart was still pounding as I realized that the rat-tat-tat that had scared me was only the teletype that had started up.

Out of curiosity I went back to see what had come over. It said, "Chicago . . . 11:03 . . . Chicago policeman killed trying to rescue two boys fallen from pier on Lake Michigan . . . Officer previously awarded two medals for outstanding bravery on duty. Bodies of three victims not yet recovered . . ."

It was eerie to watch the paper move up against the stationary carriage of the machine, to watch the sudden small snips of letters imprint themselves on the paper. Eerie, and hypnotizing, so that I stayed to see whether there would be anything to follow.

Nothing happened for a minute, and I began to walk away when the staccato sound started up again. I watched each little character fall down, leave its mark, and go on. It was like getting a message from Mars—except the dateline said, "Chicago, Illinois." After that I was no longer aware of the type or the words being formed or even of the machine itself. The item said, ". . . 11:05 . . . Dismissal of five more executives of The Kimberly Publishing Company announced by Kimberly Board of Directors. Included in mass dismissal are Robert Land, Chicago Manager . . . Maxwell Price in San Francisco . . . Alan Black in New Orleans . . . Steve Lunderman, head of Kimberly Corporate Promotion . . . Peter Greiff in London office . . ." Five in one swoop, and all of them, Greiff and Lunderman, and Black, virtually Bob Moshier's whole team, people he had brought in, good, bad or indifferent, but now part of the losing team (of which there seemed to be two) and therefore losers themselves.

"And what about—?"

The words had barely started in my mind, when suddenly the machine spit out one finishing item: "Also dismissed was Thomas R. Redford, publisher of Today magazine . . ."

twenty-five

IMMEDIATELY AFTER LUNCH, MARK POST sent a memo to each person who worked for *Today*. Every copy was personally delivered by his secretary, who placed it carefully on each one's desk as though it might explode.

The memo was typical Mark Post. It said: "I just wanted each of you on the team to know as soon as possible—and to know directly from me—that I had no idea of this nifty game of 'All fall down' that management has seen fit to have us play. I would like you to know, and believe, that the dismissal of Tom Redford was nothing I knew anything about before you did. It was as much of a surprise (and a shock) to me as it was to you. In many ways Tom was the greatest champion that *Today* ever had. His methods might have been a little unorthodox, but then you knew 'ol' buddy' as well as I did—and he did have his own way of handling things.

"I want you to know, as I have said before, that the news of Tom's dismissal was as shaking to me as it was to you. I also want you to know that I have just this minute left off on a phone call to the esteemed chairman of the Board at Kimberly. And I can promise you one thing. It's pretty hard to get through to some of those tradition-barnacled types. But I now have an absolute promise that never again will a change, be it hiring or firing, be made at Kimberly without my knowledge of it first. And I promise on all counts that that knowledge, as soon as it becomes mine, will become yours.

"I would assume that a new publisher will be announced at *Today* in the near future. I want you all to know that—whoever that person is—you will be told before the information is released publicly. So, as they say at Con Ed, 'Bear with us while the necessary improvements are being made. Dig we must . . .' " It was signed, "Mark."

As copies of the memo were read, the pitch of hysteria grew higher and higher. Salesmen congregated in each other's offices, secretaries drew together in little whispering clots, people drifted up and down the corridors. Blending together, all the low, whispering, conjecturing voices were like a buzzing thread being pulled slowly through your brain.

The strangest part of it was that whenever you met someone, you almost automatically stopped to figure out whether their jobs were ones that would be affected by the firings that had been made. The first time this occurred to me was when I walked down the hall and saw Jane Perry packing Tom Redford's things into a carton.

"Who are you going to work for now?" I asked bluntly.

She looked up at me coolly and said, "Who knows? Nobody, I guess." She even laughed a little.

"Oh," I said stupidly. "Have you—did you speak with Tom?"

"Yes. He's going to come in this afternoon, I think. He said he had a few things to clean up. I'm going to ship all his personal files to his home."

"Where are you going to look for another job?"

She laughed. "You are a worrier, aren't you? I don't really know if I'm out of one as yet."

Among the things she was packing away were two paperweights that used to sit on his desk. I hesitated, and then I said, "Do you think—could I have one of these?"

"They're just old merchandising pieces that *Today* did once."

"I know—I mean—I'd like it for a souvenir."

She stopped packing and stood up and looked directly at me and said quietly, "You really like him, don't you?"

"Not exactly—I mean—well, I guess maybe what I mean is—I don't know . . ."

"Well—that's pretty definite."

"What I mean is," I said, trying to find my way, "I know he's a nut, but he seems concerned about *people*."

"You mean he's concerned about Tom Redford."

"No," I said, floundering. "Didn't he give me a job when I didn't have one—and—and what about Jack Sheehan?"

"And what about Pete Larsen?" she said quietly. And then, "And what about the fact that you came here recommended by practically the top man in the company—and what about the fact that he's got such a hang-up about Jack Sheehan because his own father was a lush that he would probably have killed that damn sweet fool in his fanatic personal need to 'save' him . . ."

After a moment I said, "I take it you don't like Tom Redford very much."

"Yes," she said quietly, "as a matter of fact I do. I think he can be charming and sweet and maybe even, as you say, have a kind of feeling for people. God knows, he's better than a Brad Rusk, and he's nice to work for in an insane kind of way, and sometimes I even enjoy his jokes, but . . ." She hesitated for a moment, and then she said, "I really don't know you well enough to say this—but I'm going to anyway. Aren't you a little old to keep making judgments about people out of your belly instead of your brain?"

I didn't know what to say to that. She picked up the paperweight and handed it to me.

"Here," she said, "take it. Only remember, it's made out of lucite. Not that it isn't handsome. Only don't try to convince yourself it's pure crystal."

I left the office, suddenly realizing that this girl whom I had summarized merely as "cool," was much more than that.

So what if she needed to put Tom Redford down? Maybe she

just saw him in a different light than I did. In a way, I thought, it was almost funny—her needing to "protect" me. As though you could pick Redford out to be the villain of this whole conglomerate mess . . .

I passed Mark Post's office. After the slick braggadocio of his memo he looked extra-quiet—*diminished,* I think is the word I mean. And I wondered, could it just be that Redford knew more than he realized when he had summed him up? *"The trouble with Post is he's never been to the fight"?* Could it be that, never before having had to make an important moral decision, and having been seduced by his own latent ambition into an act of betrayal (what *else* could Clip's Central Park wooing of him have been about?) he would find—even if they did move him into Redford's old office and he got exactly what he came to New York to get—that he had paid too high a price for what he had bought, and that from day to day, as you went past his office, you would have the impression of his getting littler and littler behind the big, shiny publisher's desk that would grow bigger and bigger?

I was near the elevators when one of the car doors opened and Redford got off.

"How are you, honey?" he said, putting his arm around my shoulder.

"Fine," I said, "how are you?"

"Never better, honey—never better."

We walked down the hallway toward his old office, his arm still around my shoulder.

"You think some of my old friends on *Today* be wanting to wish me goodbye?"

"Sure," I said, "I guess so."

It was a long corridor to his office, and a weird thing was happening as we walked. About twenty feet before we reached any of the offices, among them Post's and West's, the door would close abruptly—sometimes when we were even closer than that. I didn't quite grasp it immediately, except that I had registered the fact that both Post's and West's secretaries were just in the act of putting down their phones a moment before we went past. And then I realized: word was out that Redford was on the premises.

For a moment I wondered whether Redford would challenge this in any way. I had visions of him bursting into both their offices and absolutely demolishing them. Instead he just said good morning to West's and Post's secretaries, "Morning, Miss Maryanne. . . Miss Sue," bowing, smiling his charming best. "Everybody certainly seems pretty busy around here today. It's a truly inspirin' sight."

We had reached his office by now. Jane Perry was on her knees packing his things.

"Hi, honey—how'd you like it if you and me started the plushiest cat house this side of New Orleans?"

She laughed and said, "It sounds like a great idea, Mr. Redford. In what exact role do you see me functioning in this enterprise?"

He laughed, but then I saw his face change as he walked into the office and the full impact of the situation took effect. Jane had taken his pictures down from the wall, and I saw him stop for a full minute, looking at the Jim Canfield ad in *The New York Times* that he'd had specially framed.

I started to go, and he turned, looked at my face and laughed out loud.

"You sure are a gloomy-looking old lady," he said, in the way only he could use and get away with. A way that I realized I would be missing.

I wasn't sure of what I wanted to say. Finally I said, "I'm sorry," and put out my hand.

He shook hands with me, and then in his own way, typically, he laughed and said, "Don't be, honey. Don't you see, it's all so simple . . . the way it always is in the long run. We took it away from somebody—and somebody took it away from us." It was quite a summation.

But then, instead of leaving me quickly, as would have been natural for him, he stayed there, holding my hand, almost awkwardly. He cleared his throat and finally said, "Paula."

I held my breath; it was one of the few times I could remember him ever calling me by my name.

"Paula," he said again, and then, quickly, "I want you to know that I appreciate the suggestions you made—trying to help me and all. And I want you to know they weren't bad—my telling them all about what I was planning on doing for the magazine and all. I used them," he said, talking faster now, almost as though he had to convince himself of something. "Maybe," he said, "maybe I even went a little overboard with using them, but—but, well—you understand—there was a lot at stake."

Then he didn't say anything, and to break up the awkwardness that had come—that I couldn't understand—I found myself saying, "Yes— Yes, of course, I understand."

He seemed released then, as though he had gotten over something that had to be done.

"Well—so long," I said again, and left quickly, and walked back to my office. Jack Sheehan was sitting at his desk, staring, as usual, down at the keys. I said, "Tom Redford's here."

Jack laughed a little cynically. "I know. I saw the doors closing as he came down the hallway. It's funny," he said, "I couldn't help thinking how much it was like a leper's bell warning all the 'pure' people that he was coming. Actually it was the same thing. Redford just happens to have contracted the most deadly of modern-day diseases—failure. Deadly—and highly contagious."

"What do you suppose he'll do now?" I said.

Jack laughed. "Get an even better job. They always seem to."

"How do you mean?"

"People like Tom Redford always go on to even better, higher-paying jobs. They thrive on being fired. You watch and see. Redford—Rusk—Moshier—especially Moshier—it's an interesting phenomenon. They use all their failures just as though they were stepping-stones. The only one I'm not sure of is Wolffe. In some way he seemed cannier than the rest. But I'm not sure he's got the same survival power."

Neither of us said anything for a while. I had not tried to reach Dave since that morning. Thinking about him again, wondering why he had not called, I felt a sense of complete loneliness, like drowning in black water. For a full minute my hand moved slowly toward the phone and back again and then again toward the phone, and back.

At that second the phone rang. It was Clip West asking me if I'd come into his office when I had a minute.

I said to Jack, "Do you believe in guilt by association?"

He looked up.

"That was Clip West," I said. "Redford's old buddy, remember? I don't suppose you'd want somebody like me around after you'd done a great job of killing your buddy. He wants to see me. I can only assume it'll be an official request to turn in my skate key."

"Maybe not," Jack said. "I wouldn't assume anything—considering the cast of characters we've got going here."

"Well . . ." I started out of the room. "Wish me luck—or something."

"Don't be so sure," Jack said, and then, "And—hey—whatever it is, don't let it throw you—O.K.?"

"I promise," I said.

Clip West's door was still closed, but his secretary—Pete's former secretary—said to me, "You can go right in. He's alone."

I opened the door and Clip West said, "Hi—I wanted to talk to you about something. Sit down. Make yourself comfortable."

I picked what seemed like the easiest chair to get out of, sat down, and waited.

"I just wanted to tell you that Mark Post and I have some real excitin' plans for promoting this magazine. First of all I was over to my buddy Ted Monger's house last night—and I tell you he's got some plans for changing this magazine that you're going to flip over."

I blinked and sat up a little straighter in my chair. "For one thing he's going to slant it a whole sight more to women. We be having more features on fashion—and food—and a whole slew of other things that ladies are interested in. What I'd love to have you think about is doing a real classy presentation on women in relation to *Today*. Something you be tailor-made to do."

And all of a sudden I understood something—a whole way of thinking that I had never dreamed about. *Of course*—Clip West never even *thought* of my leaving. As far as he was concerned I was someone who worked for him.

To a Clip West your loyalties automatically shifted to whoever was winning—the only allegiance you ever really had was to yourself. Yourself as part of the victorious team. As for this particular Clip West, in his own corrupt way, his thinking was as pure as anything you could imagine. He was in charge and

I worked for him. Of course Pete Larsen might be there for a little while, but he'd be leaving very shortly now. And nobody'd even have to ask him. And as for somebody named Tom Redford —there *had* been a buddy of his once . . . No telling what he might be doing right now. As for me, what if Tom *had* hired me? I worked for *him* now. Maybe he even got his jollies out of that.

Clip West— As Orin Kreedel had said, the shark who killed without malice.

He was still talking, but I hadn't even heard the last few sentences. All I knew was that he was standing up now and I supposed that the meeting was over. "I'm going to need me every bit of help," he was saying, "and it's nice to know I got me some professional people to lean on." And I thought, "It's just his way of saying: 'I'm top nigger now, and don't you forget it.' "

I walked out of West's office in a state of semi-shock. I felt the way you do right after you fall; you're too numbed then. The real pain will come later.

I went into Jane Perry's office. She wasn't there. Her desk was still covered with Redford's papers and correspondence. Knowing Jane, I knew she would file each and every one of them with exactly the same care she would have used if Redford hadn't already been fired.

I was about to leave when one of the papers caught my eye. It was a carbon of a memo:

To: Members of the Board of The Kimberly Publishing Company
From: Tom Redford, Publisher of *Today*

Almost without realizing it, my eye fastened on one of the last paragraphs:

"In conclusion I would like to say that I believe it is also highly indicative of the unorthodox, unbusinesslike ways of Mr. Wolffe and Mr. Rusk that, for some time now, Mr. Wolffe has been involved in an intimate relationship with one of *Today*'s employees —said employee being paid as a member of my staff in the function of a sales promotion writer. The lack of discretion shown by Mr. Wolffe in formulating this relationship is, I believe, more than sufficient evidence that Mr. Wolffe——"

"Hi—looking for me?"

I swung around, feeling the blood rush out of my head and the darkness begin. For a second I could only vaguely make out Jane standing in front of me, looking at me.

Willing myself not to fall down, I desperately managed to say, "Oh, no. I was just passing."

"Why don't you sit down for a minute. I was just——"

But I was already halfway down the hall, stumbling blindly toward my office, the cold, empty darkness of the room folding around me, the tears running hot, as I dropped my head down to the desk, crying wildly into the cold, burning cup of my hands.

twenty-six

"PAULA . . ."

The first thing that came to my mind was that I was sitting with my back toward the door. Whoever it was that was standing in the doorway wouldn't be able to see my face.

I took two deep swallows and made an elaborate pretense of sneezing three times in succession. That gave me time enough to grab at the box of Kleenex in my top drawer. The sneezing jag would cover up for the soggy mess my eyes and face were probably in.

"Paula—have you got a minute?"

I took one last deep swallow, made a final swipe at my eyes with the Kleenex and turned around.

It was Mark Post. He was standing tentatively in the doorway, sort of peering over his glasses at me.

"Paula—could you come into my office? Have you got the time? I thought we might chat for a minute."

I said, "Yes," in as calm a voice as I could manage, and he stood aside to let me go in front of him. I entered his office warily, even more so when he hurried to pull up a chair for me next to his desk, all the while making nervous little half-gestures in what I assumed was an effort to make me comfortable.

When he had managed to convince himself that I was definitely and "comfortably" sitting down, he hurried around the desk, took his seat, and then said nothing.

He made a sort of arch out of his hands, touching the tops of his fingers to each other, resting his elbows on the desk. Once every five seconds the shock that I had felt when I read Tom Redford's memo flooded over me as painfully as it had the first time.

Finally, he cleared his throat nervously and said, "Paula—Paula, I thought we should have a chat, what with all that's happened in the past few days. Now that this insane organization seems to be settled down again—anyway, God, I sincerely hope so!—I think we should get our bearings so that we can go back to our jobs without any sense of insecurity about them.

"Of course," he said, "nobody likes the mess that's been going on for the past few days. It's never very edifying to see anybody's dirty laundry, but in a way, you know, maybe there's some truth to the old saying that things work themselves out for the best in the end."

I wondered what "things" he was talking about, and then, almost as though I had heard it, I could imagine the conversation that had gone on between him and Clip West that weird day in Central Park.

"It's not just a matter of you making a better publisher than Redford," I could imagine West saying, "it's a matter of Moshier and his whole team being a *threat* to the magazine. With you as publisher and my buddy Ted Monger as editor, and me—well, me helping in whatever little ways I can—you're gonna make *Today* a magazine that Jim Canfield would be proud of having started. You'd be a great publisher, Mark—I just got a feeling in my bones you got it in you to be one of the really great publishers of all time . . ."

I could almost hear the seduction word for word. Because I was imagining how that scene had gone, I lost Post's last few sentences. When I came into focus I heard him say, "And that's why we, Clip and I—we want to make sure you're staying—we want to make sure you weren't considering anything else."

"Well, to tell you the truth, I hadn't thought about it much."

"We—Clip and I—we don't want you to feel uncomfortable or anything like that just because it was Tom Redford who hired you originally."

I hadn't felt particularly uncomfortable "or anything like that" until he mentioned it, but now that he did I could feel the little hairs on the back of my neck standing up.

I didn't say anything and he continued with almost painful and studied "sincerity." "Seriously, Paula, we—Clip and I—we want you to stay on here. Anyway it'll be nicer for you now. It was always Redford who never liked your work."

For a moment I was sure I had heard crooked. It had to be a joke—*didn't it have to be a joke?*—except that Post was saying it again, "I tell you Paula—and I tell you the truth—it was Redford who never liked your work." And then, "Isn't that right, Clip? Wasn't it Tom who was so unfair about Paula's work?"

I looked up and there he was, standing in the doorway, Clip West.

"Isn't that right," Mark Post said again. And West said, "Beg pardon? I don't think I heard."

"I was just telling Paula how it was really Tom Redford who didn't like her work."

West laughed. "Well, what do you expect from that loony character?" Among other things, evidently, Clip West had managed to perfect the art of never giving a straight answer.

But—with equal compulsion—Mark Post evidently had never learned how to leave well enough alone.

"I mean it, Paula, it was Redford who didn't like your work," he repeated. "He was an insane man. You understand, don't you?"

He waited for me to answer, not realizing he had asked the most explosive of all questions. Finally I said quietly, "I think so."

He was so pleased with my having answered, with his having "convinced" me, that he didn't know enough to drop it right there. Instead he went on, committing the cardinal of all salesman's sins—over-selling.

"Now we—Clip and I, that is—have some very special projects that we think you'd be perfect for. We want to give you every chance that Redford never did."

I was about to say something—I'm not sure what—when Clip suddenly picked up the conversation. It was almost as though, sensing where the dialogue might go, he cut into it quickly, short-circuiting anything I might have been about to say.

"I've already been talking to Paula about the fancy kind of Today's Woman presentation I thought she'd be great for . . ."

"I'd like to say something now—" But that was as far as I got, because—practically like nudging me out of the way—Clip interrupted.

"Heck—" he said, grinning sort of sheepishly, his "just folks" routine, "I wasn't even going to say anything, but I tell you something Tom Redford did the very first day I came here on the job.

"Tom called me at home that same night. He told me I had to fire you."

Suddenly I felt sick to my stomach. Not that I'd forgotten the memo of Tom's that I'd seen . . . not that I didn't know he was a kook, wild, living by a sort of "code of the street"—half-crazy, inconsistent, apt to do anything.

"I'm sorry," I said quietly, "but I don't want to play the game."

Post pulled off his glasses and said in a sort of squeaky voice, "What game is that?" Then it was Clip—edges of steel showing through all the soft southern treacle—who said, without a trace of an accent, "What game were you talking about?"

I took a deep swallow and then I said quietly, "Where I come from they call it Kick the Corpse."

There was stone silence and then Post, in his tiny, betrayed voice said, "Paula—I don't think I understand . . ."

Finally West spoke. "I understand you been with this company a long time."

"Are you threatening me?"

He said, "I take it from what you said that you don't want to work here any more."

I grinned. It was a dirty grin. "Would you want me to, now?"

He was about to answer me. I don't know what he was about

to say, and I never would know, because at exactly that moment Mark Post's secretary came to the doorway and mumbled hesitantly, "Excuse me . . . I'm sorry to disturb you, but I'm supposed to tell you there's a meeting in the conference room on the fourth floor."

"Me?" Post asked, in a dazed and confused voice.

"Everybody," she said, "everybody that works for *Today*."

"When is it?" Clip asked. And she said, "Right now, I think. His secretary said right away."

"Whose secretary?" I could see the confusion building in Post's eyes. He looked as though he'd like to go to sleep and then maybe everything would go away.

"Mr. Jefferson's," she answered. "She said he was the Chairman of the Board."

"Oh," Mark Post said. It was the tiniest sound I had ever heard.

Clip West moved toward the door. "I guess I better get my jacket." For a moment the three of us just looked at each other. Then, without saying anything, Clip left the room, and I stood there for a minute, and then I left, too.

When I got back to my office my secretary, her voice full of awe and excitement, said, "Oh, Miss Jericho . . . I couldn't find you. I'm supposed to tell you. There's a meeting——"

"I know," I said. "Thank you."

"They said it was for everybody on *Today*. I guess that means me, too."

"Yes—yes, I guess it does." For a second I found myself patting her on the shoulder, like somebody's grandmother.

The conference room on the fourth floor was a huge, dreary place with no windows, dominated by two oblong mahogany tables pushed together to make one long slab. The walls were painted a kind of sick yellow.

Most of the people from *Today* were there when I arrived. It was startling to see how many of them there were. Almost nobody had anything to say. Here and there people lit cigarettes, or played with loose clips they had found on the table, or doodled on notepads. The secretaries looked uncomfortable, as though they felt it was wrong for them to be at a meeting that included their bosses. All except one long-haired blonde who sat with her knees crossed to her navel, as though she expected to be photographed. Every once in a while someone coughed nervously or yawned. There were no conversations as such.

Suddenly the door was opened and Ted Monger and Clip West entered together. Monger, as usual, was wearing his thin, slit-mouthed grin.

Clip, just to make sure everybody knew what was what, walked in with his arm around Monger's shoulder. The new team, the gesture said. Clear the way. Watch our dust. Join up or go down. "Shouldn't there be cheerleaders for this rah-rah meeting?" Monger said. "That's the trouble with these Chicago types, they never know procedures."

Clip laughed as though it was the funniest thing he had ever heard in his life. Nobody else laughed, but I think they were beginning to get the picture. Rudolph Hecht was there, too, looking very bored, but not enough to conceal the fact that he was also looking scared.

Suddenly Monger said, "Come on, Paula, you're the girl with the inside scoop. What's the meeting for? They're not about to make you the new editor, are they?" I didn't say anything. All I did was try to ape his mocking, thin-lipped grin.

"Speaking of our fair motherland, Chicago, Illinois, I heard a pretty wild story about it the other day," Monger said. "It seems this old Irish pastor was walking around his church. It was after services . . . no one else was there . . ."

But there wasn't time to finish the story. Because at that point the door opened and Arthur Jefferson, flanked by Miss Lorraine Clapper and Mr. Graves (the little man with the black book), entered the room. Immediately there was complete silence. Graves and Clapper, one on each side of the old man, walked directly to the head of the long table. They both took seats, leaving the old man in full command.

For almost a full minute Arthur Jefferson just stood there; then he said, "I would assume that I don't have to tell you that those of us in Chicago have always been particularly proud of your magazine, *Today*. It represents the best. Always has. *Today*'s like buying a Rolls Royce automobile. Or sailing on the *Queen Mary*—or staying at that Plaza Hotel here in New York. Or smoking them three-dollar cigars that come in their own glass tubes. It's the best. That's why we never stinted on expenses. We never cut corners. Ernest Hemingway—eh, what's his name?—Faulkner—John Steinbeck—all of them high-priced people had pieces in *Today*.

"And all of you people here pretty much always do a good job in selling and promoting the book, too.

"I think Ted Monger's doing a pretty good job. The same goes for Mark Post and all the rest of you. I think your new promotion director—eh, I forgot the name just now—anyway, I think he's got some first-rate ideas for promoting a high-class book like *Today* . . ."

Out of the corner of my eye I could see Clip nudging Monger; he was grinning like they'd just made him the first southern saint. As for Monger, I could see him puffing up like one of those prickly, sickly white blowfish they sometimes hang in the window in fish stores.

After a minute Jefferson went on. "I know with all the firecrackers going off the past few days, you're probably worrying about what's going to happen next. Like, for instance, who's going to be the new publisher . . ."

Mark Post was staring at the floor; I wondered whether he was rehearsing his little acceptance speech. Rudolph Hecht was making notes; I figured he was roughing out the press release that would announce Post's being made publisher. In a way you could feel the atmosphere of the room loosening a little all around you.

Jefferson was quiet for a moment. Then he said, "I know you're all probably wondering about what other changes may be going to take place." He stopped and looked at Graves.

After a second he said, "Maybe the best way of settling your minds about the whole affair at *Today* would be to read you from a statement that'll be released to the press immediately at the conclusion of this meeting."

Slowly Arthur Jefferson took a folded sheet of paper out of his pocket, and slowly he opened it flat, and slowly he began to read.

"As of today, December fourteenth, the Kimberly Publishing Company announced news of the dissolving of one of its properties, namely its general features magazine, *Today*." A sound went through the room. It was sort of like a long, drawn-out breath of disbelief.

Jefferson let the pause build for a minute, then he went on. "Started in 1930 by the late James Canfield, the magazine quickly skyrocketed both in revenue and in prestige. The decision for the magazine's closing at this time, as described by Arthur Jefferson, chairman of the Board of Kimberly Publishing, was—quote: for purposes of tightening and strengthening the company's other publications.

" '*Today* has become, so to speak, obsolete,' said Mr. Jefferson. 'In a market of specialized publications it was just a matter of facing facts.' " Jefferson paused just long enough to catch his breath and then went on.

"If possible, the Kimberly Publishing Company announced, it will try, wherever feasible, to assimilate the services of as many of *Today*'s staff as possible. Of course this will not be feasible on higher levels, but it is believed that a good proportion of its secretarial staff and others in positions that are not executive can be absorbed elsewhere into the company.

"The company, established around the turn of the century, will continue to publish its other magazines, among them, *Woman's World*, *Ladies Fair*, and *The Review*.

He read it completely straight, without one single comment or inflection anywhere. Now, having finished, he took off his glasses, put them back into his pocket, folded the sheet of paper from which he had read, and said flatly: "I am sorry, ladies and gentlemen. But those are the facts. As I said in the beginning, it's like the *Queen Mary* all over again. I guess we all know what happened there. Times change; audiences change.

"Even the big writers *Today* ran—where are they? Most of them dead—either that or drunks. And where are the three-dollar cigar smokers? where are the people who traveled in first-class leisure aboard the *Queen Mary?* where is the audience that once thought *Today* was the classiest magazine going? I'll tell you where they are—they're gone. And just because they are gone is why it wouldn't make any sense still trying to put together a magazine for them. Not if you want to make a profit. And, like President Zachary Taylor once said, 'I ain't in this line of work to get popular.'

"Well," he said, "I guess that's it. I'm sure you top-notch executives on this magazine won't have too much trouble finding jobs elsewhere. As for the subsidiary help, the secretaries and clerks and such, I'm sure personnel will be able to fit you in someplace or other on one of the other Kimberly magazines.

"Well," he announced, "I guess that does it." At which point he—and Miss Clapper and Mr. Graves—departed from the room.

For at least a minute after they had left no one said a word or budged an inch. The first one to move was Ted Monger. Looking suddenly very ill, he quickly left the room. Clip West stayed where he was. His face seemed to be frozen into absolute shock and disbelief.

Then the salesmen started to move. The numbness wore off; you could hear the panic in their voices. "Hey, Al, what about that? Dissolved? Jesus, he didn't say dissolved, did he? Not *Today*. Did he say that?" Two of the secretaries were crying; I wasn't exactly sure why. The blonde one stayed right where she was with her legs crossed; she might be right about the photographers coming, at that.

I had to get away from the whole thing as fast as possible. I went to my office to get my coat.

I pulled on my gloves. I didn't know where I was going. I'd have to come back to settle things—but it was really all settled already. In the meantime I just had to get away. I took the steps instead of the elevator. It was a long trip down to the main floor, but at least I could feel the comfort of my feet moving—moving and moving—away—away and quickly.

twenty-seven

I KNEW THAT I COULDN'T GO HOME YET. There were things that had to be done. But I needed to get away from there for a while. It was cold in the street, damp, full of more snow. I headed over toward Central Park. It was chilly, bleak, almost deserted. Completely dazed, and in a numb kind of way, desolate, I tried to sort out the truth of everything that had happened.

As far as my own job was concerned, the whole situation was almost funny. My great last stand—my virginal speaking-out—and I hadn't even gotten past the first sacrificial sentence. It was funnier still when you realized that all the while there hadn't even *been* a job to sacrifice.

Or you could look at it another way. Since I was obviously not an executive, there was every chance that I could, as the good Mr. Jefferson had pointed out, "be assimilated elsewhere in the company." Except that, as I remembered Tom Redford's memo, "*for some time now, Mr. Wolffe has been involved in an intimate relationship,*" everything else seemed unreal and unimportant.

I had to speak to Dave. In my mind I kept seeing the picture of him that had been in the newspaper—the cold, bitter picture of someone who had been so badly hurt that he might never be healed.

It was so long—why hadn't he called me? (I didn't even dare think of the fact that he might know about Tom's memo.) How could he possibly not know how much I wanted to be of some help to him now? I felt a need to touch him, to tell him—to tell him what? I didn't know, only whatever could help.

I passed the old weeping willow, huge and mournful, that I could remember passing ever since I was a child. A wave of such absolute sadness and loss swept over me that I stumbled and almost fell. It was as though all the peace I had ever known in this place was stripped away, like something I had forfeited my right to enjoy.

Just outside the park I could see one of those small, glassed-in phone booths, and I rushed toward it. I didn't know what I wanted to say exactly—only that what had happened didn't matter, wasn't important.

Even though it seemed almost useless to call his office, only a gesture, a hoping, I decided to try it anyway. Frantic, desolate, I dialed the number, holding my breath, not daring to breathe or hope, only waiting as the phone rang directly . . . once . . . twice . . . three times——

"Hello." His voice sounded cold, far away.

"Dave," I said. "I've been trying to find you . . ."

There was no answer—none at all. "It's me—Paula." And then, compulsively, afraid of the silence that might still follow, "How are you? I've been trying to reach you. Are you all right?"

"Yes," he said. "I'm fine." And then nothing.

"Are you all right?" I said. "Is there anything I can do?"

"No—no, I'm *fine.*" It was cool, distant, polite—but that's all.

"I was sorry," I said, fumbling, awkward. "I was sorry about what happened . . ." There was no answer. "I suppose you know that they folded *Today* . . ."

"Yes," he said. "So I heard."

It was like being caught in the middle of a lonely dream, and not being able to wake up. I didn't know who I was or who I was talking to. Not really.

"Dave—are you sure you're all right?"

"Yes, of course."

Out of a desperate need, I said, "I have to see you. Is it all right if I come up to your office for a minute?"

There was a long emptiness. For a moment I thought the connection had been broken, and then he said, "I'm only going to be here about fifteen minutes. I have an appointment."

"I'll come right away," I said. "It's not far. I'll take a cab."

There was no answer, and I said, "All right? Will you wait for me? Dave?"

Finally he said, "I won't be here long."

The lobby was deserted when I got to Kimberly. I took an empty elevator to the executive floor and tried not to see the little lost groups of people standing around. I had to keep myself from running the length of the hallway to Dave's office at the end. The rooms just before his were completely empty except for a secretary who was packing some files away into a carton.

Just before I got to Dave's door I found myself stopping, and then—desperate to see him—I half ran into the room.

He was taking a Maillol down from the wall. It was the one with the woman's hands raised over her head. I remembered once he had told me that Maillol was his favorite artist. It occurred to me that I hadn't realized that the drawings in the office actually belonged to him.

He turned around, still holding the frame, as he heard me come into the room.

"Dave—" It seemed all the breath I had was just enough to make the one word.

"Oh—hi," he said, almost coolly, almost blithely, and the breath in my throat actually ached until I realized how hurt he must be to have to make this much of an effort to hide it.

I moved further into the room, my steps slow, awkward. I waited for him actually to *see* me, but he was involved with the bookshelves now, piling the books into a carton.

"Dave . . . How are you?"

"Fine." He looked at the title on one of the bindings, considered it for a second and then tossed it into the wastebasket.

"Dave—I had to see you. I mean—I had to explain."

"I've only got about five minutes," he said quietly. "You could have said whatever it was on the phone."

The room started to darken around me. I took a deep breath.

"Dave," I said—not knowing I was going to say it—"about Tom's memo. I saw it——"

For a minute he didn't say anything or even look at me. Then he turned around and stared at me for a minute. Then he turned back to the books.

"Dave—I have to explain."

"There's no time," he said. His voice was distant, freezing. "I told you you shouldn't have come."

"Dave—we have to talk about it. When I saw that memo—what I mean is, the worst part of it was that it was true. I mean, I did talk to Tom about seeing you. But that was in the beginning—I mean—I didn't know you then."

My knees were shaking. I wanted to sit down, but I had no faith in the fact that I could actually move.

"Dave . . ."

"It doesn't really matter," he said deliberately. "Anyway, I have to go now."

I watched him as he put the last book in the carton and closed the top. He flipped open each of the desk drawers; there was nothing left. I watched him as he moved over to where his coat was hanging. He was right; it was too late.

"Well," I said, "I guess I better go too. I mean, even if I don't stay I'll have to clean up things."

I waited for him to say something, but there was no question about what I might mean about staying or not staying. No questions about anything.

Like someone trying to be heard in a wind, I said, "I've decided that I might leave Kimberly."

But it was hopeless. He was going through his wallet now, looking for something. And there was no answer. No question. Nothing.

"Well . . ." I said, knowing that after this there was no more, nothing at all, nothing else I could say. I swallowed and tried not to hear my own voice.

"Will I be seeing you soon? I mean—I wondered when——"

For a whole minute he kept going through the papers in his wallet, and then, without stopping, he said, "I'm going to Washington for a while. There's a job there I'm considering."

"Oh . . . How soon are you going?"

"I have a ticket for tomorrow morning. I'm seeing the president of the company in a few minutes. He's in New York for a conference. If it works out all right I'll bring my family out there in a week or two."

It was like somebody talking to an enemy—worse than an enemy—something closer to a stranger.

In desperation I said, "Dave, you've got to understand . . . you've got to forgive me. Dave—please—Dave, I——"

The phone rang.

He hesitated and—like someone hurt badly who determines it will not happen again—you could see him steeling himself against whatever it might be. His voice was flat as he said, "Hello. Who is it?"

Whoever it was must have asked him what he was doing because he said, "I'm on my way to the appointment right now."

I was sick with desperation. It was as though he were gone already, lost, gone; the old sense of abandonment flooded over me. Frantically, I leaned over the desk and scribbled on the pad in front of him, PLEASE . . .

He never looked up at me. Instead, he hesitated for a second and then as I waited, bewildered, not understanding, he put the receiver down, reached over to a small box-like thing at the side of his desk, and flipped a switch.

A woman's voice flooded into the room. At first that was all I was aware of—just the sound of the woman's voice, like water washing into every corner.

". . . we'll wait dinner then. Anyway, I have to pick up Barbara from her dancing class. You won't be late, will you? Listen . . ."

I felt myself start to get sick. The voice went on, worried—a little preoccupied—wifely——

"I was talking to Liz Dempsey next door. She's got a brother in Washington. She says there are a lot of nice houses . . ."

The voice stopped for a moment, and then, larger-than-lifesize, grotesque and real both, it said, "Is anybody with you? I thought you had to go to a meeting. Why am I doing all the talking?"

Only then did he look at me. He picked up the receiver and, still looking at me, he said into the phone, "It's all right, hon. It's just someone from work."

I watched him as he held the receiver to his ear for a moment, and then slowly, deliberately, replaced it on the stand.

". . . for your family. I mean, well, make sure that they'll make arrangements for the house. And schools. Oh, I have to tell you, Dave—you don't have to worry about the kids. I mean about the job or anything. They don't even care about your not being at Kimberly anymore . . . Oh, listen, on the way home maybe you'll stop and bring Barbara some strawberry ice cream —three A's on her report card for this month. Oh, and if you think of it—for Richard—some little something just so he won't get jealous . . ."

Somehow I had found my way to the door. The voice was going on. "And listen—Dave, listen—don't worry or anything foolish like that. I mean—well, you have a home and a family, and we love you and what else matters if——"

Once I got to the street, I started to run, but the streets were patched with ice and I almost fell. After half a block I passed the cafeteria where everybody always went. I started inside to get some coffee, but halfway to the counter I felt the tears starting to pour out of my eyes. Desperately pretending I was having a fit of sneezing, I found my way back to the tiny ladies' room at the far end of the cafeteria.

Thank God it was empty. I stumbled into the single john and locked the door and sat down and started to cry. For a while I couldn't stop. I knew I had to get out of there but I couldn't stop crying. Cried, cried, came apart like a rain-rotted rag . . . Where had I read that?

"Dave—is anybody with you?"
"Just someone from work . . ."

Except that just then, blessedly, I thought I heard someone else about to come into the room, and the awareness at least made it possible for me to cut the edges of my crying.

I came out of the john and soaked four paper towels under the cold water faucet and held them to my eyes.

"Dave—is anybody with you?"
"Just someone from work . . ."

I started to cry again, and then I cupped my hands and let the cold water fill them, and then I dipped my face into the cup of my hands.

After a while I stopped crying. It was then that I realized that that was what you did. You stopped crying. And you washed your face. And maybe you took an Anacin or something. And then, when you could, you opened the little door, and you came out smiling.

I put on fresh lipstick and straightened my coat. Everything was in order. As I put my hand out for the knob I heard it again: *Just somebody from work . . .* It occurred to me that I would probably be hearing it for quite some time.

I swallowed past the razor blade in my throat and opened the door. The noise of the place was like something closed off—voices, groups, and all of them excluding you, and you had to go out into it, to plunge into it, the way you always had to when you opened a door.

I saw Orin Kreedel sitting alone at one of the tables. I started toward him and then, at the last second, discovered that I didn't have nerve enough to do so. Instead I sat at the counter and ordered a cup of coffee. It wasn't the coffee I wanted so much as I longed for a hot cup to wrap my hands around—to take comfort from.

Holding the cup, letting the steam of the coffee rise into my face, I was lulled. For a moment the life of eight years ago, isolated, isolating, stretched out its arms and was beautiful, was warm. But now, without choice, I saw it was the wish, the desire, the choice of those who lie down in snow, desiring the numbness of soft death to the pain of coldness and not knowing. The pain that blood can be in your veins when each throbbing brings intensity, brings further pain.

I was about to drink my coffee when, impulsively, almost

without choice, I picked up the cup and walked over to the table where Orin Kreedel was seated.

For a moment, reaching him, I saw the question in his eyes as he looked up, the challenge that was always and would forever be there.

"You don't remember me," I said.

And he said, "I do. Sit down."

I sat down, and for a moment neither of us said anything.

Then I said, "Have you heard what happened?"

"Yes."

"What do you think? How does it make you feel? I mean, being with it so long."

And he said, quietly, sanely, "All I know is that it's better to see a decently buried body than a rotting corpse."

I didn't say anything, and he looked at me directly and said, "Does that shock you?"

"No." Except I wasn't sure.

After a moment I said, "What are you doing now?"

"I've had a few offers," he said. "I'm going to decide about them this week." And then he said, "I have a feeling that I'd like to try my hand at editing books. Maybe after so many years with magazines I'd like to work with things that have a little more *permanence* about them . . . What about you? Are you all right?"

"I think so."

"Have you learned anything?" Even though he spoke quietly, without any particular emphasis, his eyes were inescapable.

"I think so," I said. "I'm not sure."

"That's a beginning. What are you going to do now?"

"I think I'll write for a while . . . ," I said, not having considered this decision until the question was put to me.

"Can you live that way?"

"I used to. Anyway," I said, "I remember something Mr. Canfield said to me one time I met him. He said I'd probably be shaken up by some of the things I'd get myself in the middle of. But then he said, 'Just pray you end up knowing who you are. God help you if you ever forget that.'"

"So you've finally confronted the ultimate dilemma."

I looked at him.

"The inability and the necessity to go on. And do you know?" he said. "I mean, who you are?"

"No," I said, "not really. But maybe what I do know is who I'm not."

Neither of us said anything for a while. I knew the odds were that this would probably be the only time I would ever see this man, that even this time had been by the sheerest chance.

"How long were you with *Today?*" I asked him.

"From the beginning."

"That must make it worse for you."

"I have an old saying," he told me. "I had it on my bulletin board at *Today* almost from the day I arrived. 'I am always poised for flight.' "

"How beautiful that is . . ." Because I didn't know what else to say.

And he looked at me with eyes that stripped away everything, all the corners, all the hiding places, all the things that said one thing and meant something else.

"Wait until you're forced to live it," he said. "That's the only time you'll be able to say whether it's beautiful or not." And, when I didn't answer he said, "What *about* you?"

"What about me?"

"You don't give yourself one hundred percent, do you?"

"How do you mean?"

"Don't get your Italian up," he said quietly. "And don't start conversations with, 'You don't remember me, do you?' unless you mean it."

Instantly I could feel the anger rise up in me; I could feel the striking-back bitchiness of what I would say next, except that just then—the almost-spoken fury hot on my mouth—I realized that he was right. My saying to him, "you don't remember me." It *had* been a ploy—or you could call it a protection, a kind of cop-out, a device used to shield yourself from rejection.

"I'm sorry," he said. "I mean—if you expected a *polite* conversation, I've never been very good at that. It's just that I have this—" he stopped, he smiled now as he said, "I guess the word nowadays would be hang-up. About people being one hundred percent straight. People that I care about," he said. "If you *had* been a stranger—if your statement about my not remembering you had been true, I'd have probably been disgustingly polite. It's just that with people that I do have some feeling about—I'm apt to turn vicious with anything that isn't one hundred percent straight. It's one of the reasons I don't have many friends.

"I'm sorry," he smiled. "I really don't know you well enough to be this rude. At least let me buy you another cup of coffee by way of amends." He signalled for the waitress. "Are you sure you don't want anything else?"

"No, thank you. The coffee would be fine."

The waitress came and filled my cup. Neither of us said anything for a minute afterwards, and then, not able to stop myself, although I was afraid of what response it would bring, I asked him, "What do you mean? About my not giving myself one hundred percent?"

"In a way it's unfair," he said. "You see, I know more about you than you know I do. It wasn't a matter of finding out—of asking questions. It was just that—well, let's just say something about you struck me that first day you came into the office to speak to me. I'm not even sure I know exactly what. Except that I felt in some way you weren't being one hundred percent fair with yourself—one hundred percent true, I guess is what I mean." He stopped. "Are you sure you want me to go on with this?"

"Yes," I said, afraid, but still more afraid to miss what he might say, what it was I might learn.

"Well," he said, "for one thing, you were working for the Promotion Department. Obviously you'd agreed to take the job there, and you were doing it, and probably doing it well. But you were ashamed of it. Yes, that's exactly the right word, ashamed. Am I wrong?"

"No. No, you're not wrong."

"Afterward," he said, "even though Monger might have killed any plans that would have involved our working together, our conversation wasn't anything I could just put out of my mind—well, maybe not our conversation, but something. Anyway—chalk it up to whatever reasons there might have been—it was as though what had been established that morning in my office was an—awareness of you."

He laughed. "Don't forget that at heart I'm a bastardly *nosy* individual—also, to put it a little more nicely for my own benefit, I have all the endlessly curious instincts of an editor. In small ways, as I said, without ever pursuing them, I was getting to know more and more about you.

"Once I saw you with Dave Wolffe. It was just for a moment on the other side of the street, and it might have meant nothing —it was even just a coincidence that I happened to notice it. But something told me it went more deeply than that. I don't know what—maybe the look on your face, maybe something about the way he was talking to you—anyway, whatever it was, I knew it was more than just a casual saying hello between two people who just happened to work for the same company.

"That's all it was—an incident. Except I wondered . . . You know, of course, that everyone assumed you were sleeping with Tom Redford . . ." He looked at me with a glance that could have pierced rock. "I didn't come to any firm conclusions about you one way or the other," he said. "What I did know was that you were close to him. Your face always seemed to—radiate, I guess is the corny word—when you were around him. As I said, it wasn't any business of mine. Merely the observations of an inquisitive man.

"Except one thing killed me. I had spoken to you . . . you seemed, underneath all of the messy confusion of Kimberly and Redford, pretty direct—honest. Except how could you be? Isn't there something in the Bible about no man being able to serve two masters?"

"But—I was working for both of them. I mean——"

"Please," he said, "I'm sorry. You don't have to defend yourself with me. Why should you? When I really don't even have the right to bring up the subject."

"No," I said, "please tell me. And, anyway, you'd be doing me a favor."

He looked at me for a long time. Then he said, "Actually, there isn't much more to say. I just wondered if maybe you weren't cheating yourself. If you're going to be a writer—be a writer. And I'm not saying you have to starve while you're doing it. But just *do* it, for God's sake. Even if it's only for two hours a day. Be committed to it. Be committed to something.

"And if you're going to be a writer—then get a job someplace . . . Kimberly, IBM, a shoe store—whatever it is, be committed to *that*. And the same thing goes for people. Be for them or against them. Stop pussyfooting. You can't cop out by being for *everything*. Don't you see, you'll never really know who you are until you find out exactly what it is you really stand for."

And it was true—all of it.

After a moment I said, "I envy you. I wish I had known Mr. Canfield.

In a very low voice, so that at first I wasn't sure if I had heard him right, Orin Kreedel said, "He was a man."

I looked at him.

"Let people be themselves," he said. "Canfield was a man. He was most probably the best magazine editor alive in the past hundred years. He was also almost a full-time drunk and on occasion a first-rate bastard, when he had to be."

My confusion must have shown, my disbelief, my wild, dizzying drawing away.

"Stop making Canfield into some super-sized plaster cast statue that's got 'Editor' stamped on the base. Stop making him an *absolute*, stop forcing halos down around people's heads. Is it that much of a compulsion? Do you have that much of a need to *worship?*"

I felt as though I had stepped off into a deep, dark hole, the depth of which I couldn't really judge, mainly because I couldn't stop falling. He—the questions he asked, the answers he demanded—surrounding you—were everywhere.

"Maybe even me," he said, going on, inescapable. I think you think I'm a saint. Or a genius. Or—God knows—something somebody 'made up.' How would you handle the fact that, say if we got to know each other at all—you'd have to see you were wrong? How, for instance, would you handle the realization that there are times when I have a fierce, foul temper? that I'm frequently rude? that I sometimes get sick to my stomach?

"What do you do?" he said, "what do you do with these fake statues after they're built? How do you keep from finally having to *see* them? What do you *do?* Run? It that what you do? do you run from all these 'perfect' people?"

Suddenly I was crying. But he didn't turn away. He didn't

seem embarrassed. He just waited. It was as though, however long it took, he would just wait.

"They die," I said.

"Who besides Canfield?"

"My father."

"How old were you?"

"Seven."

"And Canfield," he said. "And, God knows, maybe others . . . I told you, I have a foul disposition," he said. "And Canfield had a mistress—a woman you might not even have liked. And your father—your father was your father . . ."

I didn't say anything. I had stopped crying, but I wasn't sure when. My eyes were still wet. And my mouth.

"Aren't *people* enough?" he said. "The ones that mean something to you? Can't you—embrace us the way we are? Any other way is like holding onto a shadow. Can't you do that?"

"I don't know."

"At least try."

"I will."

"All right then," he said. "Try. Try is a beginning."

I felt a deep, long breath build in my stomach and then rise to my chest and then reach my mouth—and it was all very long, and slow, and simple. Like drowning. Or breathing on the top of a very high mountain. I wondered what time it was. I felt that I had been here for hours. Either that, or that I had just come.

"What about now?" he said. "Do you mean it when you say you're going to write—or is that just a statement with reservations, too?"

"I don't know," I said. "I think I mean it, but I don't know."

"Meaning isn't necessarily *knowing* a thing," he said. "It's doing it. It wouldn't have been worth a goddamn my just griping about what a disaster Monger was as the editor of *Today*—griping at home, to friends, to other fellow-gripers at the magazine. Unless I told Monger what I thought of him—to him—directly to his face.

"I remember, I had a secretary once—Marcia. She was a Negro girl, but very light, the kind they refer to as being 'able to pass.' One day I realized that she was wearing a special makeup—to make her look darker. And I asked her why she did that. And she said, 'Mr. Kreedel, it's because when people see me, I want them to know exactly what I am—one hundred percent.' "

After a moment he said, "Can you live on that? writing?"

"I don't know. My mother has a small pension. We used to."

"What about your expense account? And the fat-cat lunches magazine people get used to? And that Air Travel Card I expect you have? Can you get used to not having any of that anymore?"

"I don't know," I said. "I'll have to see."

"I assume you're a Catholic," he said.

And I said yes. And he shook his head, smiling a little sadly. "I don't understand," he said. "I don't know too much about your religion, but I do know that you think all of this"—his gesture indicated the cafeteria, the street outside the dusty windows, Kimberly, the world—"is for a reason, an end. Is that right?"

"Yes."

"Then why don't you live that way?"

"I don't understand. Live what way?"

"With faith," he said. "As though you really believe what you say you believe."

Neither of us said anything. He was ready to go.

"Maybe we could talk again sometime. If you'd like it."

"Yes," I said. "I would like it."

"Good, then." He stood up. "Well . . . goodbye for now. I wish you luck."

"Thank you. You too."

He left and then, a moment after he was gone, I left too.

It had started to snow again. I would go back to the office and straighten out whatever was left to straighten out. And then. And then I would do whatever I would do then.

And, standing there, I knew that all that Dave and I had ever really had was part of the terrible and beautiful excitement of the special, insane time we had both found ourselves in the middle of. In a way it had been like the intimacy of two people sharing a bomb shelter. When the bombs stop falling they are strangers again.

And in that terrible true second of vulnerability—because of just having talked to Orin Kreedel—I had to face still another truth, the fact that after two months of "Don't call Steve" scrawled at the top of every calendar page—Monday . . . Tuesday . . . Wednesday . . . Thursday . . . Friday . . . *"Don't call Steve"*—I was sick with the aloneness of being lonely. And that would be another thing I would have to learn how to live with, no matter which way anything went.

And at that moment, with the wind icy up my sleeves and down the neck of my coat, I began to find a small portion of the truth of myself. The taste of it was sharp and bitter and acid, like a worn copper penny in my mouth. It was a taste I would have to get used to: honesty. With myself. For *me*. And I hoped it would work. Here, there, Kimberly, whatever—I hoped that I would stay, or go, or do whatever it was I did, for the at-last, simple reason that finally I wanted to *grow up and be me*.

I continued to walk back slowly through the icy streets. And all of a sudden I was aware of the hypocrisy of my own beliefs—or at least what I had always believed to be my beliefs. Because—what was I doing? I had prayed to God to have this problem taken out of my hands . . . I had said, "*You* had better do something, because look at the botch job I'm making of it," and now, done, answered, granted, I felt lost, bitter, wanting it the way it had been before. And I wondered, Why do people pray at all? Because they never really want what they ask for. If people really had any faith, they'd be too afraid to pray.

Walking, half-dazed, I faced, forever and irrevocably, the fact I would have to learn to live with. A larger truth than I had ever imagined and one that I seemed to have run away from most of my life. That I was, in fact, many more people than I had known—that my motives were not, would never be, the open, obvious ones I had lived by. And grown fatly smug on.

And in many ways I hated this knowledge of myself. Because it seemed that, understanding yourself, you were never freed from the understanding of other people, too.

Tom Redford—"We took it from somebody, and somebody took it from us"—Even this wild philosophy, outrageously outrageous, I could only understand. And that understanding without choice. But merely achieved. And once achieved, an albatross, never lost. Clip West, Mark Post—it was terrible once you understood, because you could never really fight the logic of understanding—you could never again escape into the beautiful black-and-white world where everything had been marked, accurately and finally. It was terrible to understand, and for a moment all my instincts fought against it. One needs the succor of one's own blind spots, and to have them illumined—to understand deeply the weakness of Redford, the pitiableness of West, the minor tragedy of Post, the truth of Dave and myself—the fact that he thought I had betrayed him. And—the final fact—that I had . . .

Again, and even more strongly, it came to me that the most painful part of this was that there was nothing about what had happened that you couldn't understand. It was all so clear and simple. Tom—the nightmare memo. Words written out of fear. For his job, for his family, for the nightmare image of going back poor to Mississippi, for God knows what else. And Dave. Hurt. Acting out of hurt. *And* betrayal. (That was, after all, the right word, so say it.) Even Clip West . . . Once you had grasped the idea of how he had scratched his way up (up?) through who knew what odds, even his greed and conniving were something that—though it might make you sick to your stomach—still added up. So what was there left not to understand?

So that all of a sudden I wanted to yell. I wanted *not* to understand. I wanted to have all my own little biases and prejudices intact again—all I could find. I wanted to use them the same way other people did—to protect myself with, to blind myself—not to understand. God, never to have to understand.

Anyway, what did it mean—to understand? All you did was fuck yourself up understanding. It meant you couldn't blame anybody. It took away all the villains. It meant you had to bleed and puke and sweat for everybody else's hang-ups the same as if they were your own. And who the hell wanted that? Get rid of it . . . throw it away . . . burn it out . . . whatever. *Doctor, I have this problem . . . I mean, it's killing me . . .*

And standing there in the street, alone, I laughed out loud like a nut, my head thrown back. An over-age hippie type complete with sandals and beard said, "Hey, bird—you really blown your mind—right?"

I couldn't stop laughing, "You *know* it baby . . ."

There was a magazine stand outside of a branch of the Chase Manhattan Bank. There was a poster tacked to the side of the stand. It said, "Read TODAY . . . news . . . travel . . . sports . . . women's features . . . fiction . . . science news . . . and more, and more, and more, and more . . ."

In a week the sign would disappear—in a week or a month, it made very little difference which. And it would never be replaced. Anyway, not with one that said TODAY. And to most people it would matter very little. If at all. A world had died. But then worlds were dying all the time. And who cared, really?

And then, inescapably, I remembered Canfield. ". . . you picked just about the most lunatic pit in the world to muck through. But there's one thing it can teach you—if you survive. And that is to see *people* . . . And that means you, too."

It was almost what Kreedel had said.

". . . It'll probably shake you up plenty, some of the muck you're going to get in the middle of. It'll take a peasant's ankle—and honesty—not to drown in it. But that's the only thing that'll give any purpose to staying here. You think you got it?"

Remembering that, I yanked up my collar against the cold. I would go back to the office now. I would finish everything that needed to be finished. And then I would begin. I would begin again.

I wished myself luck.